W9-AUV-913

DISCARDED

THE ESSENTIAL GUIDE TO CULTIVATING MUSHROOMS

THE ESSENTIAL GUIDE TO CULTIVATING MUSHROOMS

Simple and Advanced Techniques for Growing Shiitake, Oyster, Lion's Mane, and Maitake Mushrooms at Home

STEPHEN RUSSELL

Storey Publishing

The mission of Storey Publishing is to serve our customers by publishing practical information that encourages personal independence in harmony with the environment.

Edited by Carleen Madigan and Claire Mowbray Golding
Art direction and book design by Mary Winkelman Velgos
Text production by Jennifer Jepson Smith
Indexed by Christine R. Lindemer, Boston Road Communications

Cover and interior photography by © Stacy Newgent, except: © Bon Appetit/Alamy, 55; © DP Wildlife Fungi/Alamy, 24; © GAUTHIER Stephane/SAGAPHOTO.CC/Alamy, 227; © imageBROKER/Alamy, 25; © John Greim/fotolibra.com, 143; © Stephen Russell, 16, 30, 39; © Susan Waughtal, 34 (bottom); © Tammy Venable/Alamy, 15
Illustrations by © Marjorie Leggitt, 18 and © Michael Gellatly, 89, 99
Mushroom spore prints throughout by Mary Velgos, photographed by Mars Vilaubi

© 2014 by Stephen Russell

All rights reserved. No part of this book may be reproduced without written permission from the publisher, except by a reviewer who may quote brief passages or reproduce illustrations in a review with appropriate credits; nor may any part of this book be reproduced, stored in a retrieval system, or transmitted in any form or by any means — electronic, mechanical, photocopying, recording, or other — without written permission from the publisher.

The information in this book is true and complete to the best of our knowledge. All recommendations are made without guarantee on the part of the author or Storey Publishing. The author and publisher disclaim any liability in connection with the use of this information.

Storey books are available for special premium and promotional uses and for customized editions. For further information, please call 1-800-793-9396.

Storey Publishing
210 MASS MoCA Way
North Adams, MA 01247
www.storey.com

Printed in the United States by Versa Press
10 9 8 7 6 5 4 3 2 1

LIBRARY OF CONGRESS CATALOGING-IN-PUBLICATION DATA

Russell, Stephen (Stephen D.), 1984–
The essential guide to cultivating mushrooms / by Stephen Russell.
pages cm
Includes index.
ISBN 978-1-61212-146-8 (pbk. : alk. paper)
ISBN 978-1-61212-463-6 (ebook) 1. Edible mushrooms. 2. Mushrooms. I. Title.
SB353.R87 2014
635'.8—dc23

2014015198

CONTENTS

Preface

Deciding to grow mushrooms has turned out to be one of the most rewarding experiences of my life. In many ways, it has shaped my entire lifestyle: it has enlivened my curiosity about the natural world, guided my plans for the future, and helped shape the way I view the planet. Most others I know who have been ensnared by the hobby find it equally compelling. The ultimate intention with this manuscript was to develop the book I wish existed when I first made the decision to become a cultivator.

When I began growing mushrooms a decade ago, the best references were Paul Stamets's books *The Mushroom Cultivator* and *Growing Gourmet and Medicinal Mushrooms*. They were so influential that I often referred to them as the Old Testament and the New Testament. However, while the Stamets books were extraordinary — even revolutionary for mushroom cultivators — they did not fully elaborate on many of the most beneficial methods for small-scale, at-home cultivation. They also did not address many of the unique challenges home growers often encounter.

With that in mind, I set out to produce a book of solid methods that readers could replicate for success, or use as a base for their own experimentation. Using the step-by-step methods presented here, hobbyists can successfully grow a supply of mushrooms for their home dinner tables, or fill up their basements with mushrooms to sell at weekly farmers' markets. Each chapter includes a troubleshooting section that lists common mistakes or problems that new growers are likely to encounter.

There are several things to keep in mind when reading this book. First, there is no "best" or "right" method for growing mushrooms; instead, there are many different paths to the same goal. The best method for one cultivator may be terrible for another, depending on a number of factors, including climate, family situation, cost, species desired, and the availability of materials, time, and space. I have outlined many of these considerations while discussing the various methods. Even if you are familiar with growing plants, remember that working with mushrooms requires a quite different set of processes and conditions. Some of the processes may not become entirely clear until you actually try them and watch them work. Attempting the processes is the best way to learn them.

Second, this book should be approached in a stepwise manner, as many of the basic methods and chapters are building blocks for the more advanced methods. If you attempt the more advanced methods before you have a firm grasp of the basic methods, you'll probably introduce a variety of contaminants to your product, resulting in failures that you will not have the experience to diagnose. The most successful mushroom growers take the time to learn the basics and approach the process in a methodical manner.

Experimentation is good, but only if you don't adjust too many variables at the same time. You can expect a certain amount of failure in mushroom cultivation, so if you can't narrow down the source of that failure, then you'll learn nothing from the experiment. And try not to let failures get you down. Over the course of time, I've made nearly every mistake possible, many of which are described here. Try to view them as learning experiences, and then try to avoid repeating them!

My approach to growing mushrooms is based on the science of controlled experiments: I want to replicate my successes and minimize my mistakes. That's why this book stresses indoor cultivation methods. Growing mushrooms outdoors often involves too many variables — it's unpredictable and sporadic. It is also seasonal, depending on your climate, so

you may only get a couple of harvests a year. Indoor mushroom cultivation, on the other hand, will yield consistent and predictable harvests year round. It is the gold standard of cultivation, but to be able to do it effectively requires a serious commitment of time and learning. If you can make this commitment, the end result is truly rewarding, and you will have gained a significant and marketable skill set that few people have attained.

I hope the information in this book will enrich your life in much the same way it has enriched mine.

PART 1

Basics for Beginners

CHAPTER 1

GETTING TO KNOW MUSHROOMS

People often ask me, "Why mushrooms?" They usually ask the question even more insistently when they find that I typically don't eat most of the mushrooms I grow or find in the wild. Until you get to know mushrooms, you may not realize the complexity of the mushroom world. Their world is one of endless fascination and intrigue. Mushrooms have innumerable uses and innumerable stories to tell.

Everyone knows that mushrooms can be used as food, but did you know that you can use mushrooms to make paper, paint, fabric, dyes, and hats? Did you know that there are mushrooms that glow in the dark? Are you aware that mushrooms are miniature chemical factories, and the toxins that fungi produce are used to make citric acid, birth control pills, and penicillin? Did you know that there are "zombie mushrooms" that can take over the brains of insects, force them to climb upward to the highest point possible in a tree, colonize their bodies, and grow a mushroom — all to release their spores? The more I learn about mushrooms, the more enthralled I become.

The Culture of Mushrooms

Mushrooms play many different roles in human life. One of those roles is as a cultivated crop, and that's the primary focus of this book. But as mushroom growers you should also know something about the other roles of mushrooms, because you'll certainly encounter them as you explore the intricate and diverse world of the mushroom.

Mushrooms for Food

When most people think of mushrooms, they're probably thinking of brown Portabellas or White Buttons, the common grocery store mushrooms that many of us enjoy on our salads or hamburgers.

White Buttons and Portabellas are species of the genus *Agaricus*. These are rarely grown on a small scale because their cultivation is dominated by the major mushroom farms, highly mechanized operations that mass produce mushrooms for low prices that small farms cannot hope to match. *Agaricus* species mushrooms are grown on a manure-based substrate that must be composted before it can be inoculated. This process can take from weeks to a month or more. After the composting process, the fungi are introduced to the growing area.

Agaricus species mushrooms are referred to as *secondary decomposers* because of the two-stage decomposition process that must occur before the mushrooms fruit and are harvested. The time-consuming preparation of large amounts of compost, combined with the low market price, make White Button and Portabella mushrooms low-priority species for most small growers.

Most of the mushrooms grown by hobbyists and small-scale cultivators are *primary decomposers*. Primary decomposers perform the first round of decomposition of a substrate, which is often a wood-based medium. They are the first (primary) organisms that begin to break down the cellular structure of the wood

Shiitake (*Lentinula edodes*) and Lion's Mane (*Hericium erinaceus*) are two of the gourmet mushrooms that can be cultivated indoors.

on which they are being grown. Unlike secondary decomposers, such as *Agaricus* species mushrooms, primary decomposers don't require an extended composting process to prepare the substrate.

Most of the "gourmet" mushrooms that cultivators are interested in, such as Shiitake and Oyster, are primary decomposers that break down their substrates in order to grow. Organisms like these are known as *saprophytes*. Saprophytes decompose organic material, such as the cellulose and lignin in wood or other plant material, to gain the energy they need for their biological processes. Some of the most common substrates for gourmet mushrooms are sawdust, woodchips, and straw.

Many of the most popular types of edible mushrooms, such as Morels and Chanterelles, cannot be grown by humans, so people hunt them. These

Smooth Chanterelles (*Cantharellus lateritius*) can't be cultivated, but are a popular foraged mushroom.

types of mushrooms are called *mycorrhizal* fungi (*myco* = fungi, *rhiza* = roots). Neither primary nor secondary decomposers, mycorrhizal fungi get their energy through symbiosis with trees and certain plants. The fungus increases the surface area of the roots of its host tree by several orders of magnitude, scavenging the soil for water and nutrients that were previously inaccessible to the tree. In return, the tree provides the fungus with carbon in the form of sugars generated through photosynthesis in the leaves. More than 90 percent of all land plants form a mycorrhizal symbiotic relationship with fungi.

We have not yet been able to replicate this complex biological interaction in the laboratory for most species, so none of these common wild mycorrhizal mushrooms can be grown under controlled conditions. Some of the most popular edible mycorrhizal species are outlined in the table shown here.

Common Edible Mycorrhizal Mushrooms	
King Bolete	*Boletus edulis*
Chanterelle	*Cantharellus cibarius, C. lateritius, C. appalachiensis,* etc.
Morel	*Morchella esculentoides, M. angusticeps,* etc.
Black Trumpet	*Craterellus cornucopioides, C. fallax*
Hedgehog	*Hydnum repandum, H. umbilicatum*
Bolete	Certain *Boletus* spp.
Milk Cap	Certain *Lactarius* spp.

Mushrooms for Fun

Collecting mushrooms for the dinner table is only one of many reasons to hunt mushrooms in the wild. Many people hunt them for the same reasons that others go birding or enjoy a hike in the woods. The ability to identify mushrooms in the wild is a rare talent and a way for people to connect with an important but rather mysterious part of nature.

Many states have local mushroom clubs where people can meet and take mushroom walks in the woods, learning the key characteristics used for identifying mushrooms. Organized gatherings of mushroom hunters are often referred to as "forays." The North American Mycological Association (NAMA) is the umbrella organization for all of the local mushroom clubs in the United States and Canada.

Morels (*Morchella* spp.) are another kind mushroom that many beginning foragers start with.

NAMA even formed a committee to help amateurs cultivate mushrooms as a hobby and for profit. There is an online forum at the NAMA website for cultivators, and the committee puts on cultivation events at the yearly NAMA foray. Several local mushroom clubs also have committees and events that are specifically related to mushroom cultivation.

Even if your interests are primarily cultivation, I encourage you to hike in the woods, collect wild mushroom species, and work to identify them. This is particularly easy if you have a mushroom club near your home, with professionals available to advise you, but it's entirely possible to learn the major mushroom species in your area on your own. Most regions of the country have regional field guides available that serve as a great starting point when learning to identify mushrooms. Observing and collecting mushrooms in the wild will enhance your abilities as a well-rounded cultivator: you'll notice the mushrooms' natural environments, their growth patterns, and their ecological associations. Aside from that, it's great exercise with a built-in learning component!

Mushrooms for Health

Every year, scientists, doctors, nutritionists, and the public learn more about the medicinal benefits offered by many varieties of mushrooms. Even some of the most commonly cultivated mushrooms (e.g., Shiitake, Maitake, and Lion's Mane) have been shown to be effective in fighting many different ailments.

Reishi, also known as the "Mushroom of Immortality," can be found in the wild throughout much of the world. It has been used medicinally in China for thousands of years, and scientific research shows that Reishi could have very real health impacts in Western societies as well. One of its primary biological benefits is its enhancement of immune system response. It has also been shown to help with stress, blood pressure, blood sugar levels, and cholesterol. And it's a mushroom that can be grown effectively at home, indoors or out.

Reishi (*Ganoderma lucidum*), also known as the "mushroom of immortality," is found in the wild but can also be grown effectively at home.

Mushrooms as Hallucinogens

No conversation about growing mushrooms is complete without a discussion of the role of active mushrooms in our society. *Active mushrooms* are mushrooms that contain hallucinogenic compounds such as psilocybin. Although illegal in most states, it's hard to deny that active mushrooms are a significant part of mushroom culture, and the hallucinogenic properties pique many people's interest when they consider mushroom cultivation.

When I mention mushrooms, most people joke about one of two subjects — either that mushrooms will kill you, or that they'll make you hallucinate. While there are surely mushrooms that can do both, the vast majority of mushrooms do not belong in either category. And although the vast majority of mushroom cultivation methods are the same no matter what species you choose, this book is not a guide to the cultivation of species that contain psilocybin. The focus of this book is on nonactive edible species that are grown as a source of food.

If you're looking for more information about active species, and you're in a place where their cultivation is legal, I point you to the online mushroom forums mentioned on page 228.

Mushrooms as a Business

As a result of the continuing rise of the sustainability movement, the local foods movement, and the desire for a healthy lifestyle, mushrooms are now a product with a ready market. Specialty mushrooms, once sold only in gourmet and health food stores, are now available in most major grocery stores. This means that there's potential for mushrooms to be grown as a commercial commodity in almost every region of the country. And because mushrooms have a very short shelf life, grocers and restaurants are always looking for local sources rather than having them shipped from long distances.

Both cultivated and wild mushrooms are high-value crops for farmers' markets.

All the skills presented in this book should be viewed as prerequisites for anyone contemplating commercial mushroom production. Most of the commercial growers I know started on a small scale in their own homes using these same methods. Once you've mastered these methods, you should have the knowledge base to easily expand to sales at the commercial level.

Mushrooms in Research

Mycology is the field of mushroom science. Mycology means, literally, "the study of fungi," and its breadth encompasses all of the mushroom world, not to mention countless molds, rusts, and other fungi. An estimated 1.5 million species of fungi are thought to exist, of which only 10 percent are known to science. There are many left to discover, but there are very few mycologists in the United States or around the world, which means there's every possibility for a passionate researcher to make a mark in the field.

If you're an ecologist, you could consider researching why we can't grow mycorrhizal mushrooms. If you're involved in biochemistry, medicine, or industry, you probably already understand that mushrooms produce thousands of compounds that could benefit human health and civilization. Isolating or synthesizing just one of these unknown compounds could have important beneficial effects.

The Cultivation Process

No matter what species of mushroom you want to cultivate, the life cycles will be essentially the same, so you'll follow the same basic set of procedures.

Spore Germination

Mushrooms reproduce by forming spores. Spores germinate when they come in contact with a substrate that has the right mixture of nutrients, moisture, and temperature. When they land on a suitable substrate, they begin to germinate and produce new cells that extend outward, forming filamentous structures called *hyphae*. Each viable spore that encounters the right surface will send out its own hyphae, and when two compatible types of hyphae encounter each other, they join up and exchange genetic material. At this point, the rate of hyphal growth increases, and a large, interwoven mass of hyphae, called *mycelium*, is formed. Expanding this mass of mushroom mycelium, while keeping it uncontaminated, is the primary goal of any cultivator.

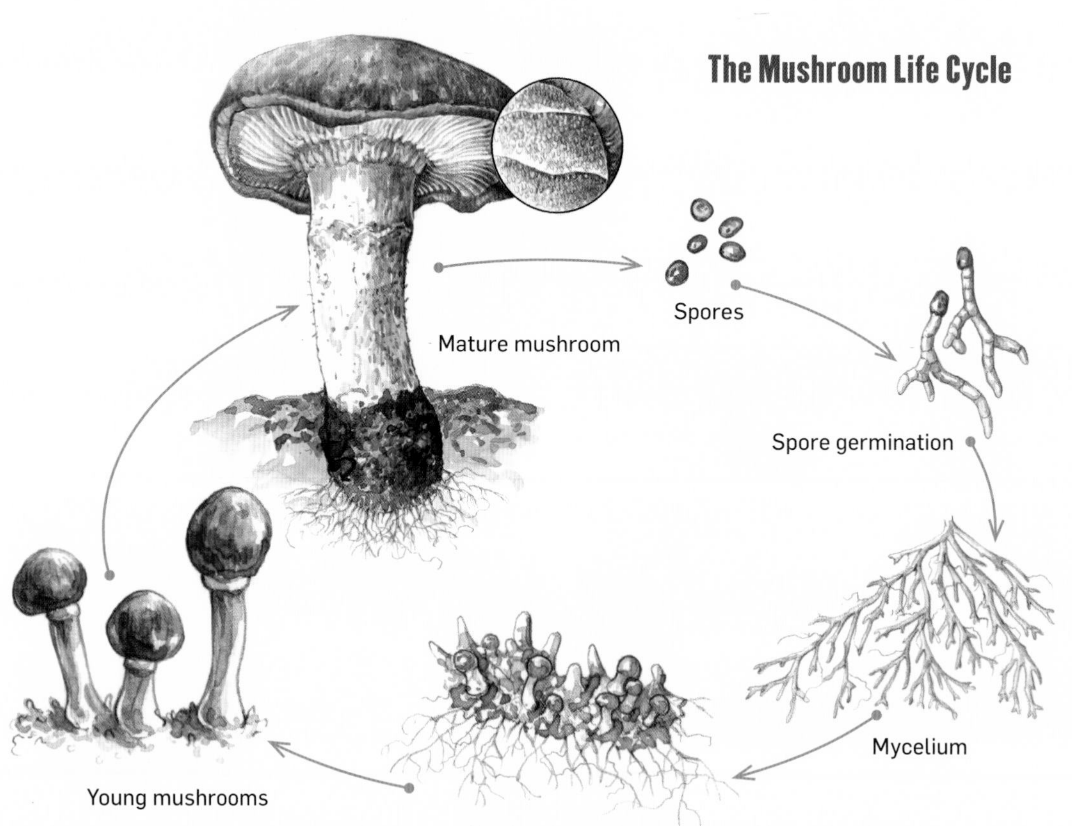

Shiitake mushroom mycelium begins its expansion through oak sawdust.

Mycelium Growth

The more mycelium you can reliably produce, the more mushrooms you'll be able to grow. Using an analogy from the plant world, think of growing mycelium as the vegetative phase of the growth cycle. During this phase, the mycelium is trying to colonize and decompose any material that it can. It's exuding enzymes (which promote decomposition) onto the surface of the substrate, and the nutrients that are made available are absorbed back into the mycelial network. This mycelial network will continue to expand until it runs out of a fresh nutrient source to colonize or until it encounters another competitive organism, such as a contaminant in the culture.

For cultivation, the mycelium is usually grown on a grain-based substrate in canning jars during the vegetative phase. Some common examples of substrates include a brown rice flour–vermiculite mixture in half-pint canning jars, or hydrated rye grain in quart canning jars. Any substrate must be sterilized using a pressure cooker before it's inoculated with mycelium. This kills off competitive organisms that may contaminate your culture and inhibit further growth.

Increasing the Mycelium

To keep increasing the size of the mycelial mass, mycelial networks can be transferred from one colonization vessel to another larger one. For example, you could inoculate a small jar of grain with an initial culture and allow that jar to become fully colonized. You could then break up that colony and transfer it to six or seven fresh jars of grain under sterile conditions. You'd allow each of these freshly transferred jars to colonize as well. Finally, the contents of each of these jars could be transferred to a spawn bag containing 5 pounds (2.3 kg) of grain. In the end, you'll have turned 2 cc of spore fluid or liquid culture into 35 pounds (16 kg) of grain spawn, capable of producing many pounds of mushrooms.

Transferring Mycelium to a Fruiting Substrate

Before mushrooms can begin to form, the spawn must be transferred to its fruiting substrate. Beginners growing small amounts of mushrooms often inoculate the fruiting substrate directly, so no transfer is necessary. This greatly reduces the chance of contamination. For growing larger amounts of many common edible mushrooms, like Shiitake and Maitake, this fruiting substrate is wood based, so you must transfer your colonized mycelium to a bag of sterilized sawdust supplemented with wheat bran. For Oyster mushrooms, straw is the most common fruiting substrate.

Fruiting

The final phase is the fruiting phase, when the mushrooms actually begin to form. Once you have successfully colonized the amount of mycelium you need, you'll want mushrooms to begin to form on your fruiting substrate. You accomplish this by creating the right environmental conditions for fruiting, which typically involves adding light, increasing the humidity, and decreasing the temperature and the levels of carbon dioxide (CO_2) by introducing fresh air. This set of actions will cause *primordia* (small, immature mushrooms) to appear on the surface of the mycelium. Once primordia begin to appear, the substrate is said to be *pinning*, or beginning to grow actual mushroom fruitbodies. Primordia can become fully grown mushrooms, but at this point, the process is in its most fragile state, and special care must be taken to maintain consistent environmental conditions. This is the only way to ensure that the mushrooms form properly.

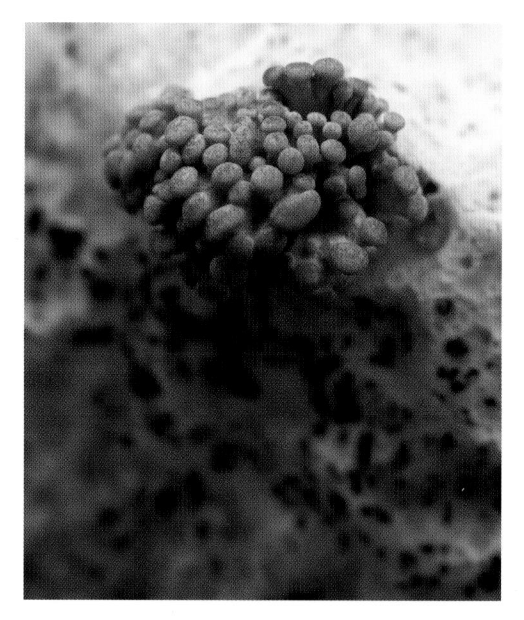

The growth of mushroom fruitbodies begins with a process called *pinning.* These primordia will eventually become mature mushrooms.

Common Cultivated Species

When selecting a species of mushroom to cultivate, it's best to start with one of these commonly cultivated species of saprophytic mushrooms. Cultures of these fungi, available from online stores or culture banks, have generally been selected for their ease of cultivation and high yields.

OYSTER (*Pleurotus ostreatus*)

Origin: Found worldwide

Decomposition: Primary; saprophytic

Description: Often white or brown. Grows directly from wood. Stem is attached laterally, to the side of the cap.

Where to find: Fallen hardwoods, North America

Ease of cultivation: Easy

Growing season: Year-round

Growth characteristics: Very aggressive

Spawn substrates: Grains (rye, millet, popcorn, birdseed); sawdust

Spawn containers: Canning jars or spawn bags

Spawn incubation temperature: 75°F (24°C)

Fruiting substrates: Wood, straw, coffee grounds, cornstalks, cottonseed hulls

Fruiting containers: 6–14" (15–36 cm) plastic tubing; plastic bags

Fruiting temperature: 65–70°F (18–21°C)

Fruiting time: 2–3 weeks after spawning

Fruiting humidity: 90–95%

Fruiting notes: Can be fruited from many different container types; most commonly grown from plastic tubing or sawdust blocks. For how to prepare straw and fruit oyster columns using plastic tubing, see chapter 13.

Harvest: Pick clusters when spores begin to drop.

Post-harvest (actions): Multiple flushes are common; leave substrate in fruiting environment.

Other strains: Pink, blue, and yellow varieties

SHIITAKE (*Lentinula edodes*)

Alternate names: Black Forest Mushroom, Oakwood Mushroom, Chinese Black Mushroom

Origin: Japan, China, Korea

Decomposition: Primary; saprophytic

Description: Brown cap and stem, with white hairs on cap surface. Meaty flavor.

Where to find: Cultivated only; not found in the wild in North America

Ease of cultivation: Easy–medium

Growing season: Spring to fall with plug spawn; year-round with sawdust bags

Growth characteristics: White, cottony mycelium browns as it ages.

Spawn substrates: Grains (rye, millet, popcorn, birdseed); sawdust

Spawn containers: Quart jars to large spawn bags

Spawn incubation temperature: 70–80°F (21–27°C)

Fruiting substrates: Wood: indoors on supplemented sawdust blocks; outdoors on hardwood logs

Fruiting containers: None

Fruiting temperature: 60–70°F (16–21°C)

Fruiting time: 30–60 days from inoculation

Fruiting humidity: 70–85%

Fruiting notes: Mycelium browns as it ages; colonizes in 2–3 weeks. Once full colonization is achieved, expose block to light while still in bag to start fruiting process. Shortly after fruiting begins, mycelium should begin to form an irregular surface, slightly warty, with some warts growing irregularly to the size of cotton balls. When irregular growth is clearly defined, open bag and initiate fruiting conditions. Some growers wait until they see pins begin to form. Pinning process may take 45 days after sawdust is initially spawned.

Harvest: Gently twist mushrooms at base when cap is fully open.

Post-harvest: Cold-shock dunk of Shiitake blocks may be beneficial for subsequent harvests. After first harvest, submerge blocks in cold water for 12–24 hours. Cold shocking stimulates pin formation; submersion in water adds moisture for the fruitbodies in next flush.

Other strains: Many commercial varieties available.

LION'S MANE (*Hericium erinaceus*)

Alternate names: Bearded Tooth

Origin: Commonly found on downed hardwood logs across North America

Decomposition: Primary; saprophytic

Description: Large white mass with white spines, beardlike appearance; texture similar to cooked lobster

Where to find: Fallen hardwood logs east of the Mississippi; rarely sold commercially

Ease of cultivation: Easy

Growing season: Late summer/fall

Growth characteristics: Thin, white, wispy mycelium

Spawn substrates: Grains (rye, millet, popcorn, birdseed); sawdust

Spawn containers: Quart jars to large spawn bags

Spawn incubation temperature: 70–75°F (21–24°C)

Fruiting substrates: Wood: indoors on supplemented sawdust blocks; outdoors on hardwood logs

Fruiting containers: Polypropylene bags

Fruiting temperature: 65–75°F (18–24°C)

Fruiting time: 15–30 days from inoculation

Fruiting humidity: 90%

Fruiting notes: Begin by colonizing quarts; spawn these jars into supplemented sawdust bags as the fruiting substrate. After bags colonize for about 2 weeks, cut 2–4 small slits into side of bag. Cut slits even if sawdust block does not appear to be fully colonized, as mycelium of this species can be wispy and does not fill out as thickly as other species. No need to remove colonized sawdust block from spawn bag. Place bag under normal fruiting conditions after slits are made.

Harvest: Twist mushroom off at base when spines begin to elongate.

Post-harvest: Multiple flushes are common; leave substrate in fruiting environment, but cut holes in new locations for the fruits to emerge from.

MAITAKE (*Grifola frondosa*)

Alternate names: Hen of the Woods, Sheep's Head

Origin: Commonly found in North America and Japan

Decomposition: Primary, saprophytic

Description: Rosette of grayish petals coming from a common stalk

Where to find: Base of oak trees east of the Mississippi

Ease of cultivation: Difficult

Growing season: Fall

Growth characteristics: Mycelium is cottony and irregular

Spawn substrates: Grains (rye, millet, popcorn, birdseed); sawdust

Spawn containers: Quart jars to large spawn bags

Spawn incubation temperature: 70–75°F (21–24°C)

Fruiting substrates: Wood: indoors on supplemented sawdust blocks; outdoors on buried hardwood logs

Fruiting containers: Polypropylene spawn bags

Fruiting temperature: 50–70°F (10–21°C)

Fruiting time: 30–45 days from inoculation to pinning; another 30–45 days for frond development

Fruiting humidity: 75–85%

Fruiting notes: Likes fruiting temperatures a bit lower than most other species. After supplemented sawdust spawn bag is fully colonized, leave bag upright and intact for several more weeks, but begin exposing to light. At 30–45 days, grayish clumps of primordial should begin to form on upper surface of block. When clumps are about 2" (5.1 cm) high (which may take 2 weeks), cut a slit in bag just above sawdust block, and introduce fruiting conditions. This allows fresh air to enter, and should encourage frond development. After 3 days, remove upper half of plastic bag completely, leaving lower half of spawn bag around sawdust block to help retain moisture. Getting to this point may take 60 days or more. Formation of fruitbodies will take an additional 2–3 weeks.

Harvest: When fronds are fully developed and edges begin to darken, remove fruitbody at base.

Post-harvest: Multiple flushes difficult with this species, rarely providing more than one

CHAPTER 2

BASIC GROWING OPTIONS

So you think you want to grow mushrooms? Wonderful! Before you make any significant investments, I recommend that you get to know the growing process slowly, step by step. The easiest way to do that is by purchasing ready-made components.

Ready-to-Grow Kits

The first option you should explore is a ready-to-grow mushroom kit. These kits are available online for most of the edible species, such as Oyster, Shiitake, Lion's Mane, and Maitake. They consist of a mass of mushroom mycelium that is ready to enter the fruiting stage. The main advantage of these kits is that you don't need to worry about learning sterile procedures, building a growroom, or maintaining cultures. Advanced procedures like those are best left for growers who have decided to commit to the time and expense of truly learning the hobby. Until you've made that commitment, kits are the perfect starting point.

Ready-to-grow kits allow you to learn the needs of the mushrooms you want to grow without much investment. The manufacturers of these kits have spent weeks or months preparing the kit before you purchase it. By the time the kit arrives, you're left with only a few simple steps to complete in the cultivation process. This allows you to become efficient at these steps before spending months preparing a substrate of your own from scratch.

The ready-to-grow kits for gourmet mushrooms usually consist of a 5-pound (2.3 kg) block of sawdust or straw in a

Note the expansion of the mycelium as it grows. These bags of fruiting substrate are shown at 1 day, 1 week, and 1 month after inoculation.

polypropylene filter patch bag. The substrate has been colonized with the mycelium of your chosen mushroom species. You will often see some letters and numbers on the filter patch of the kit. This usually indicates the species of mycelium in the bag and the date the bag was inoculated.

The only equipment you should need for growing these kits are a humidity tent (usually included), a plate, and a spray bottle. Instructions will vary slightly for each species, but for Shiitake, the process is as simple as removing the plastic bag, setting the colonized block on a plate or other base, covering the block loosely with a humidity tent, and watering the block with a spray bottle several times a day.

You can purchase many different species as ready-to-grow kits, and your first mushroom harvest can be expected within as little as 2 weeks. Kits cost between $20 and $30, and you should expect a success rate of nearly 100 percent for easy-to-grow species such as Shiitake and Oyster mushrooms. Some mushrooms, such as Maitake and Reishi, are a little harder to grow, and the fruit bodies will take longer — possibly as long as 2 months — to mature. Although harvests vary by species, you can expect to harvest between 1 and 2 pounds (0.5 and 1.0 kg) of mushrooms from each kit.

The instructions for these kits are simple and straightforward. You don't need any special tools, lighting, or skills to be able to generate your own mushrooms. The drawback is that the kits are expensive relative to the cost of making your own. It wouldn't be economically feasible to use premade blocks from another company to grow mushrooms to sell at a farmers' market, for example, unless you received a great bulk price for the blocks. In most cases, you would need to learn the methods and procedures outlined in the upcoming sections of this book to grow enough mushrooms to sell.

The instructions will vary if you're growing another species, and those instructions should be included in any kit that you purchase. It's best to start following the kit's instructions as soon as possible after receiving it. The ability of mushroom mycelium to fruit in a kit will degrade with time.

A Kit to Avoid

Morel kits (*Morchella* spp.) are one of the few types of kits you should avoid. These kits are heavily advertised throughout the Midwest, and most likely other places as well, but there are very few people who are successful with kits of this species. While these kits contain living mushroom mycelium, it's very difficult to get the kit to produce mushrooms.

Growing Shiitake Mushrooms from a Kit

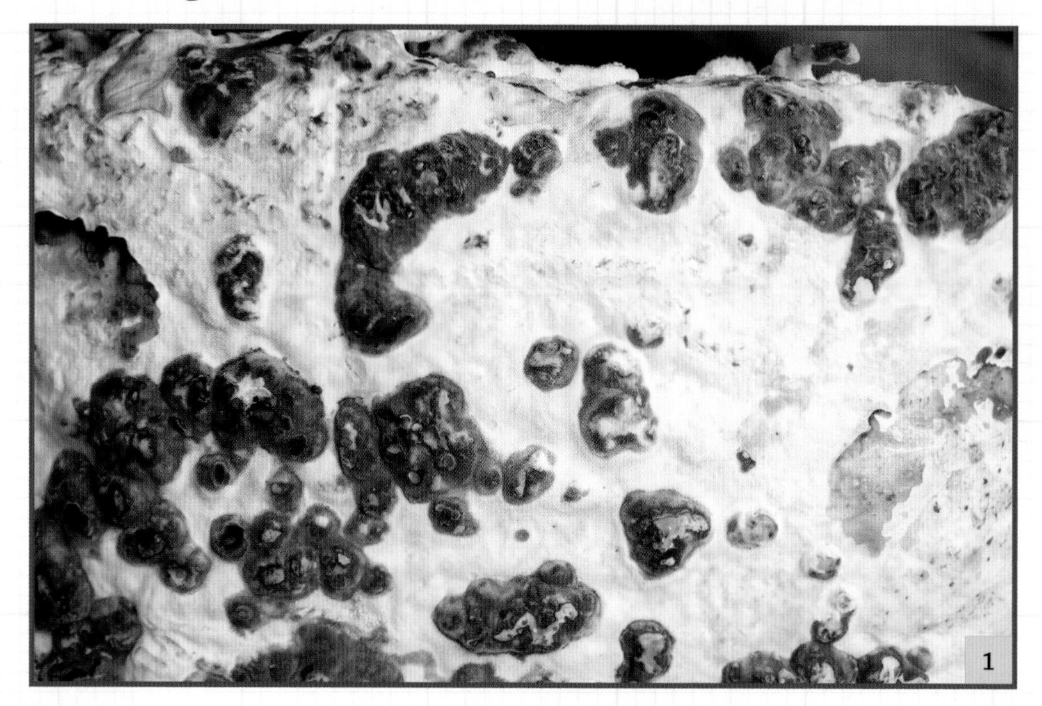

1 **Wait for buttons.** Start by looking at the date on the filter of the plastic bag that contains your block. If it hasn't yet been 45 days since the date listed, do not remove the block from the bag. Place the bag, with the block still inside and sealed, in an area where it will receive indirect sunlight. Leave it there until 45 days from the listed date, then continue on to the next step. Some people like to leave the block sealed in the bag a little longer, until tiny buttons begin to form on the surface of the block. These are the initial formations of your Shiitake mushrooms. This may take another week or two. If you begin to see these buttons forming before the 45-day period ends, go ahead and continue to the next step.

2a

2b

3

2 **Remove the block.** Begin by washing your hands. Remove the white block from the plastic bag and place it on a dinner plate. By removing the block from the bag, you're exposing the mushrooms to a sudden influx of fresh air (oxygen) for the first time. This is one of the triggers that encourages the mushrooms to begin fruiting. Another important trigger is the decrease in CO_2 levels.

3 **Assemble a humidity tent.** A humidity tent will keep your block from drying out. Plastic shopping bags are widely available and make excellent humidity tents. Loosely drape the plastic bag over the mushroom block without allowing it to touch the block. One way to support the bag is to stick two or three chopsticks into the block. This action won't harm your block, and will keep the bag off the sides of the block.

4 **Find a good location.** The best location is a spot in the house where the block will not be disturbed by people, animals, or air currents. You also want to put it in a place that will receive some ambient light, but not direct sunlight. Shiitake mushrooms, like many gourmet species, require some light to grow; the mushrooms will not form properly in complete darkness. The optimum temperature for Shiitake is between 55 and 65°F (13 and 18°C). They will grow fine at a normal room temperature in the low 70s, but if you have a cooler location, it may be best to locate your block there.

5 **Mist daily.** To care for your mushrooms as they fruit, all you need to do is remove the humidity tent and mist the block with a common spray bottle. Ideally, this should be done two or three times a day. Once you see mushrooms beginning to form, avoid spraying them directly. Instead, just spray around them. Replace the humidity tent after each misting.

6 **Harvest.** Mushrooms will begin to form about 1 or 2 weeks after you remove the original outer plastic bag. The mushrooms will begin as small bumps on the surface of your block, and will soon turn brown and form the typical mushroom stem and cap. You may see the white surface of your block begin to turn into a brown crust as the block begins to age. This is a normal part of the Shiitake growth process.

Your Shiitake mushrooms can be harvested as the cap separates from the stem and begins to flatten out. The ideal time to harvest is before the edges of the cap start to turn upward. Remove your mushrooms by cutting them at the mushroom's base, or gently twisting the base until it breaks free.

It should be possible to get three or four *flushes*, or harvests, from your Shiitake block, potentially more. After your first harvest, remove the humidity tent and stop misting. The block will become dormant for about a week and will dry out slightly during this time.

After a week, soak your block in a pail of cold water overnight. While tap water works fine in most localities, spring water or rainwater is better. Avoid distilled or chlorinated water. The block will float, so you'll have to weigh it down with something to keep it submerged. If any green mold appears on the tough brown skin of the block, just gently rinse it off in the sink. Repeat this soaking process after every flush of mushrooms until the block is no longer producing or is only producing a few mushrooms.

For Experienced Growers

Buying a kit is also a much cheaper way to acquire a mycelial culture from a given company. This option is usually much cheaper than buying a commercial culture on a petri dish. Consider saving a chunk of mycelium from a purchased kit and growing it out on agar (see chapter 11) to save the strain for your library. If you choose to obtain cultures this way, be sure to check the terms and conditions of the purchase with the company you ordered from to ensure you're not violating the terms of sale.

Premade Spawn

Another premade option that you can purchase is called *spawn*. Spawn is a mass of mycelium that has been grown out with the intention of eventually transferring it to another substrate. This expands the amount of mycelium you have for fruiting in the future. Spawn can come in many forms. *Grain masters* usually come as colonized rye or millet in quart or half-gallon canning jars. These master cultures are transferred to other grain jars (see chapter 6) or to bag cultures under sterile conditions. Bag cultures of *grain spawn* (see chapter 12) or *sawdust spawn* (see chapter 9) are also commonly available as premade spawn. These come in polypropylene filter patch bags that arrive with 5 pounds (2.3 kg) of a grain, such as rye or millet, which has been colonized with the mycelium you choose. Bags of grain spawn are usually transferred to sawdust blocks, or, as in the case of Oyster spawn, pasteurized straw (see chapter 13). Other simple substrates such as toilet paper rolls (see page 33) or coffee grounds can also be used with premade Oyster spawn.

Similar to grain spawn in bags, sawdust spawn arrives with 5 pounds (2.3 kg) of sawdust that can be broken up and transferred to other media, such as enriched sawdust blocks (see chapter 9). The most common use of sawdust spawn for beginners is to inoculate logs that are fruited outdoors (see page 37) or by using Oyster sawdust spawn on pasteurized straw, as mentioned above. The process for

Growing spawn — simply a large quantity of mycelium — on grain or sawdust enables you to produce more mushrooms.

Shiitake mycelium will naturally brown as it ages.

making your own sawdust spawn can be found in chapter 9.

A final type of spawn, *plug spawn*, is used primarily for outdoor cultivation. The base substrate for plug spawn is 5/16" × 1" (8 mm × 2.5 cm) spiral-grooved birch dowel pins that have been colonized with mushroom mycelium. These are used for the cultivation of logs outdoors. The process is fairly simple. Holes are drilled into the logs and the plugs are hammered in. The mycelium then colonizes the log and mushrooms fruit from the wood. A full description of this process can be found on page 37.

Starting with Oyster Mushrooms

There are several beginner-friendly methods designed specifically for Oyster mushrooms. Even if Oysters are not your chosen variety, this section will help you build your skill set and develop the thought processes required for mushroom cultivation in general. This method also works for several other species, so there is room for experimentation, but it works particularly well with Oyster mushrooms.

Oyster mushrooms are known to grow on an exceptionally broad range of substrates. Essentially, Oysters will grow on anything made of cellulose, including most paper products: books, newspapers, even toilet paper rolls. Their growth is very aggressive, and they will rapidly colonize many of the substrates they are presented with. I've even seen Oyster mushrooms fruiting from concrete, and many cultivators have seen them grow from wood in or near the growroom. This is why beginners often choose to order Oyster grain spawn or sawdust spawn to begin their adventures in cultivation. Because of their extraordinarily aggressive growth habits, they are one of the few species I recommend beginning with. You can expect high rates of success with whatever process or method you choose for fruiting.

There are a number of beginner-friendly methods to use with Oyster spawn. These include transferring to straw; outdoor inoculating logs; or the easiest starting point, inoculating common cellulosic material such as toilet paper rolls. The following method outlines Oyster spawn inoculated on toilet paper rolls, but I have also fruited Oyster mushrooms in a similar fashion using shredded newspaper, hemp rope, and tissues in several different tissue box sizes. In the latter case, the Oyster mushrooms will colonize not only the tissue inside, but the entire cardboard tissue box.

Novel fruiting methods like these are not ideal for commercial Oyster mushroom production, but they do have a significant purpose. These experiments help build your arsenal of knowledge about the needs of mushrooms, as well as their growth habits. Once you know that you can successfully fruit Oyster mushrooms on a variety of substrates using purchased spawn, you can be sure that the spawn you spend time making yourself in the future will not go to waste.

Growing Oyster Mushrooms on Toilet Paper Rolls

Materials

- Oyster spawn (either grain spawn or sawdust spawn)
- Large filter patch bags
- Toilet paper rolls (Generic white rolls of toilet paper work best. Avoid kinds that have color printing, added scent, or lotions.)

1

2

1 **Dip the rolls.** Bring a pot of water to a boil. Either dip your rolls of toilet paper into the boiling water using tongs, or pour the boiling water over the rolls of toilet paper in a sink. This is a "soft" (i.e., not strictly complete) pasteurization of the roll. An oven rack placed over the sink makes a good makeshift rack to hold the rolls.

2 **Inoculate the rolls.** Allow the toilet paper roll to cool on a small plate. Crumble up your Oyster spawn and fill the interior of your toilet paper roll with it. I generally also dust the exterior of the roll with some spawn. Enclose the roll and plate in a filter patch spawn bag. Use a rubber band at the top of the bag to close it.

3 **Air and mist.** Open the bag several times a day to air it out and mist it with a spray bottle. Keep the sides of the plastic bag moist by making them the primary target of the spray bottle. Try to avoid water pooling in the bottom of the bag. If you need to, you can go without misting for a day or two without negative consequences.

4 **Harvest.** Your roll should be fully colonized in 2 weeks or less if you're using Oyster mushrooms. Your fruits should come a week or two after full colonization. To harvest, simply grasp the mushroom at the base of the fruit body and gently twist it off with your fingers. You can sometimes get two or three flushes from this method as long as the roll doesn't dry out. Subsequent flushes should start around 1 week after the previous harvest.

Growing Shiitakes from Plug Spawn

Plug spawn is the easiest and most common way for people to grow their own mushrooms outdoors. Plug spawn kits, consisting of spiral-grooved dowel pins that have been colonized by mycelium, can be purchased from many mushroom spawn companies. This method only works with wood decomposing species, so you'll be limited to varieties such as Shiitake, Oyster, Maitake, Lion's Mane, and Reishi.

The process is very simple. You provide a log, drill holes into the log, and hammer the wooden dowels into the holes. At this point, the log should be placed in a shady location and allowed to colonize. It generally takes 9 to 12 months before you get your first harvest of mushrooms, but you can expect to harvest mushrooms for 3 to 5 years into the future.

Most species of hardwood logs will work. If you're able to choose your hardwood species, white oak is best. It's a very dense hardwood, which decomposes more slowly than other species. This means that your log will last longer and produce more harvests than softer species of hardwoods. Other species that will work are maples, elm, sweetgum, other oaks, willow, and alder. Some tree species to avoid are conifers, cedar, hackberry, and dogwood.

Look for logs that have a nice layer of bark on them that is not chipped off or otherwise damaged or peeling. The longer the bark stays on the log, the more

Spiral-grooved dowel pins are the base substrate for plug spawn.

moisture the log will be able to maintain, and the more harvests you'll get. The bark also provides protection from other competitive fungi.

If a log has been allowed to sit out in the open air for an extended period of time, there is a high probability that it will become colonized by some of the many other types of fungi that typically break down and decompose logs in nature. Older logs are likely to contain these natural fungi, which will compete with the mushrooms you're trying to grow. Try to find logs that have been cut in the last 3 months.

The best time to harvest your logs is when the tree is dormant: anytime after the leaves fall off in the fall, but before the buds start to appear in the spring. The wood contains higher levels of nutrients and sugars at this time, which is advantageous for mushroom growth; however, you can expect success from logs cut at any time throughout the year.

Ideal logs are between 4 and 6 inches (10 and 15 cm) in diameter. If the logs are much thinner than 4 inches (10 cm), they will not hold water as well and will have a tendency to dry out. This inhibits colonization. Alternatively, if you're using logs with diameters greater than 10 or 12 inches (12 or 31 cm), the fungi will have a hard time completely colonizing the logs' interior. Larger logs are also heavier, more difficult to work with, and require many more plugs (which means more cost), without the benefit of many additional mushrooms. Growers generally prefer logs that are between 2 and 4 feet (0.6 and 1.2 m) long.

This method works well for the beginner at home, and it's the same method that's used commercially for Shiitake cultivation outdoors. A home grower may grow three or four logs, while a commercial outdoor Shiitake farm may grow three or four thousand. Regardless of the scale, the process is identical.

Growing Maitake and Reishi on Logs

These species often grow from stumps, so if you aren't inoculating into actual stumps, try to simulate the stump habitat. The logs can be buried standing upright, with one-third to one-half of the log underground. Another option is to bury the entire log 1 inch (2.54 cm) below the surface of the ground in a shady location. If you choose to inoculate a stump, find a tree that has been cut fairly recently. Just as you don't want to inoculate decaying logs, you don't want to insert plugs into a decaying stump. Inoculate stumps directly above the root zone, near the base of the stump, as well as on the top of the cutoff zone. Maitake can take from 12 to 36 months to fruit from a stump or log.

Inoculating a Log with Plug Spawn

1

2

3a 3b

1 **Hydrate your logs.** Freshly cut logs should be allowed to sit for around 2 weeks before you inoculate them with plugs. Living trees naturally produce antifungal compounds that begin to dissipate shortly after the tree has died. These compounds will inhibit the growth of your mushrooms, so it's best to wait before you inoculate a freshly cut log.

After your freshly cut logs have sat for 2 weeks, soak them in water for 24 hours before you inoculate them. This will ensure that they have enough moisture for strong mushroom growth.

2 **Prepare inoculation sites.** Using a $\frac{5}{16}$-inch (0.8 cm) drill bit, begin drilling a straight line of holes, 4 to 6 inches (10 to 15 cm) apart, down the log. These holes should be about 1½ inches (3.8 cm) deep. Use a drill bit stop to quickly and easily get the proper depth. The next line of holes should be 4 to 6 inches (10 to 15 cm) away from the first line, and should form a diamond pattern, with the new holes placed diagonally from the original holes.

3 **Insert plugs.** Inoculate the logs by hammering a plug into each of the drilled holes using a rubber mallet or a common hammer. Ideally, the tops of the plugs should be slightly below the surface of the bark.

4 **Seal the holes.** Seal the holes using melted cheese wax or beeswax. Apply a small amount of melted wax to the top of each inoculation site. This can be done using a brush, a dauber, or any other appropriate tool. The wax seals the drill site, preventing other molds and fungi from competing with your gourmet mushroom strain, and helps keep the inoculation sites from drying out.

5 **Aftercare.** After your logs are inoculated, you need to give them the best colonizing and fruiting conditions possible. Begin by finding a shaded area out of direct sunlight. Under the canopy of the woods is ideal, but any shaded area will work. Mesh shade cloth can help make sunny locations more suitable for inoculated logs. Many growers build stacks with their inoculated logs once they find the right location. If you only have a couple of logs, you can just lean them on the shady side of a barn or house, or lay them on the ground. Some species may need to be partially buried. (See Growing Maitake and Reishi on Logs, page 36.)

In most U.S. climates, normal rainfall should be sufficient to keep your logs hydrated. If there is a dry spell, however, it may be necessary to water or soak your logs again. The goal is to keep them moist. Check on your logs every couple of weeks, and if they seem dry, give them a bit of water.

Many growers cover their logs with straw or a breathable tarp during the winter months. Mold can form if you cover your logs with plastic. Another option is to bring them into the garage for the winter. Keeping them out of the freezing weather will cause less stress on the culture. Soak the logs in water for 12 to 24 hours in the spring to give them a good start for the year.

6 **Wait for fruit.** After your logs have been inoculated and placed in the perfect location, you must wait for them to fruit. If you inoculated them in the spring, you might obtain your first fruits by the fall, depending on such factors as the dimensions of the log, how often you add moisture, the weather, the temperature, etc. It may take up to a full year to get your first fruits, but you should be able to expect repeat flushes for the next several years.

Syringes

When you purchase premade spawn, you aren't truly growing your own mushrooms from scratch. Someone else has done the lab work and met all the sterilization requirements for you. If you want to take the next step in cultivation, you'll need to perform many more of the steps yourself. Using culture or spore syringes allows you to begin to take some of those steps.

The primary reason why most people fail at growing mushrooms is contamination. Growing mushrooms is a race between the culture you're trying to grow and the millions of other bacteria and molds that are also trying to multiply. Mushroom cultures usually become contaminated by a fault in some aspect of the sterile procedure explained in the next chapter. By purchasing your initial culture in syringe form, you're ensuring that it will be free from these sources of contamination, and it will give your initial attempts the best chance for success. Many people continue to purchase these sterile syringes of inoculum for the first several years they grow mushrooms.

When trying to decide between spore syringes and liquid culture syringes, always select liquid culture if it's available for the species you desire. These syringes contain cultures developed by experienced cultivators and selected based on their rates of colonization and fruiting ability. A liquid culture syringe will colonize faster than spores and give more predictable yields. The only reason to consider buying a spore syringe is if the species you desire is not available in liquid culture form.

The average price of a liquid culture syringe or a spore syringe is between $10 and $20. In future chapters, I'll explain how to make your own spore syringes and liquid culture syringes, but at this point in your cultivation education, it's a small price to pay to ensure a solid start to your first real grow.

Spore Syringes

Some mushroom species are only available as spores. A spore syringe is an easy and reliable way to transport mushroom spores in a sterile form. The product usually consists of a clear 10 or 20 cc syringe with a 16-gauge needle. The syringe has been filled with a mixture of sterile water and mushroom spores. Remember that individual spores are microscopic, so you

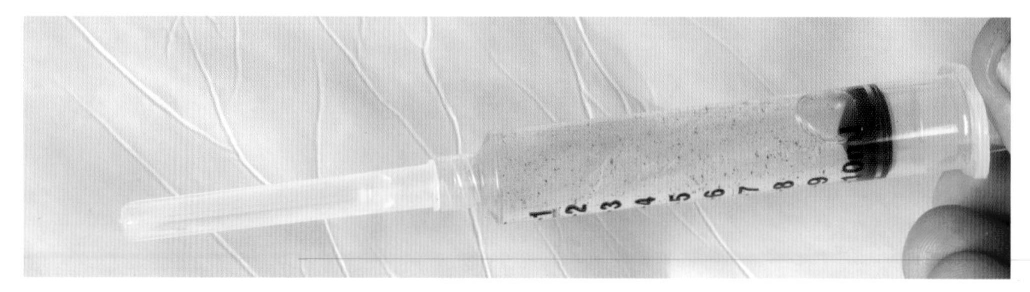

Spores are suspended in a 10 cc syringe, ready to inoculate a prepared substrate.

may not always be able to see them inside the syringe. Spores are visible only in large clusters. Thus, receiving a clear spore syringe does not necessarily mean that the syringe is faulty. The ideal spore syringe will have some small black flecks of spore-mass floating within the liquid.

Some companies offer syringes that look like black ink because of the large number of spores in the water. Some people suggest that those darker syringes allow you to use less liquid on your substrate because of the higher spore concentration, but I disagree. Regardless of the apparent number of spores in a syringe, I would still inject the substrate with the same amount of liquid. It's the liquid that allows the spores to travel more broadly across the substrate. Using less liquid from a more concentrated sporemass will limit the distribution of spores within the substrate. So when you inject less liquid, you inoculate a smaller area of the substrate. If dark syringes are the same price as normal ones, I'd purchase the dark ones, but I wouldn't pay a premium for them.

Some companies offer their syringes as "microscopy kits," which include a glass microscope slide and a cover glass. Before you order spores, check into any local laws concerning the spores of certain mushroom species. The process for making your own spore syringes can be found in chapter 4.

Liquid Culture Syringes

Liquid culture syringes are another way to transport sterile, living mushroom cultures from one place to another. The advantage of a culture in syringe form over other types, such as petri dish cultures, is that the syringes are ready to inject immediately into a jar of grain or other substrate. Because the culture is in liquid form, it disseminates throughout much of the jar right away, meaning the colonization of the jar will occur very quickly. Liquid culture syringes are commonly available for most edible species, including Shiitake and Oyster mushrooms. These common edible species would be good choices for your first experiments with liquid culture syringes.

CHAPTER 3

COMBATING CONTAMINANTS

Once you've ordered your initial syringes of spores or liquid culture, it's time to start thinking about what you'll need to do when they arrive. Virtually every procedure, from preparing your substrate to the final fruiting process, needs to be performed in conditions that are as sterile as possible. You now know that, at its core, growing mushrooms is a race between the mycelium you're trying to grow and the other organisms that will try to contaminate your culture. The best substrates for mushrooms are also the best substrates for many other types of molds and bacteria, which can and will inhibit mycelial growth. The goal is to grow out your mycelium as quickly as possible, before any competitors have a chance to take hold. To give your mycelium the best possible chance, you'll want to eliminate or minimize sources of contamination in your home, and create the most sterile working conditions you possibly can.

Keep It Clean

As an at-home mushroom grower, you should try to keep your home as clean as possible all the time. This means no dirty dishes getting moldy, no stale food in your garbage disposal, no dust on your knickknacks, all the nooks and crannies in your bathroom free of mold, and so on. All these potential sources of contaminants can destroy your hard work and cause problems well into the future.

No matter how clean your house may be, you've seen dust and contaminants in it. The dust that circulates through your home is composed of human skin cells, clothing fibers, plant pollen, and a variety of other substances, including fungal spores. These airborne mold spores are one of the primary sources of contamination in your projects.

Every home is different. Some people may have homes clean enough to allow many of these processes to be performed in the open air, including agar transfers. Others find it impossible to take a clean spore print anywhere in their home. Many factors affect home cleanliness, including the year the home was constructed, the building materials that were used, the home's history (such as flooding), the type of HVAC system, and so on.

When I begin my work in the kitchen or the *cleanroom*, the sterile area where I do my cultivation work, I think of all the dust that occurs naturally in the air. If I bring something into my sterile work area from outside, it must first be sterilized, even if it has only touched the air — as you now know, air is far from sterile. Whenever you pick up a new tool or move your hand to grab something, ask yourself if the movement you're making might create contamination. The same goes for other surfaces: if it's possible for dust to build up on a surface, you must sterilize it if you plan to work on it or touch it. For example, I always resterilize my gloves if I grab a piece of paper towel from outside my work area, if I adjust my chair, or if I scratch my face.

Develop a Cleaning Protocol

Before you start any mushroom-growing project, consider taking the following steps to clean your home. You may not have to act on all these suggestions; it depends on your individual circumstances.

Before You Start

- Clean your house throughout; mop and dust
- Shampoo carpets at least once
- Wash all rugs or consider removing them permanently
- Clean the refrigerator
- Keep the garbage disposal clean and flushed of food
- Pay special attention to nooks, crannies, and corners in bathrooms (high humidity = more chance for molds)
- Do not let trash sit indoors; keep a lid on garbage
- Remove indoor plants (soils contain undesirable fungi)
- Dust ceiling fans
- Change bed linens regularly
- Keep pet areas and beds clean
- Caulk leaky windows
- Keep shoes at the door
- Do not work in carpeted rooms, or consider laying plastic

Prepare to Work

- Shut off central air, air conditioners, fans, etc.
- Clean surfaces with a 10:1 solution of water to bleach, Lysol, isopropyl, or similar cleaner
- Shower; wash hands
- Wear clean clothes
- Work on a hard, smooth surface
- Do not work on the floor
- Run air purifiers in your work area
- Have dust masks and hairnets available for use

Ongoing

- Keep garbage disposal clear of food
- Remove expired food promptly from the refrigerator
- Vacuum regularly, but never just before you work as it kicks up particles
- Dispose of contaminated jars early
- Never open contaminated jars indoors

Know Your Enemies

When you grow mushrooms, you are attempting to create optimal conditions for the mushroom mycelium you're propagating. Unfortunately, those conditions are also optimum for many other organisms that we call contaminants. The most common contaminants in mushroom cultivation are molds (which are fungi as well) and bacteria. If these contaminants come in contact with your substrate, they'll grow fairly rapidly, often faster than your mushroom mycelium.

Problems with contamination are the primary reason people give up growing mushrooms. It can be very disheartening to have to throw away an entire batch of jars that you spent hours making up, or an entire rack of casings that you spent a month getting ready. Advanced cultivators experience less contamination than novices because they are able to spot potential contamination early and remedy the situation before it can spread. Learning to spot potential contaminants early is the best way to ensure success in your projects and prevent future problems.

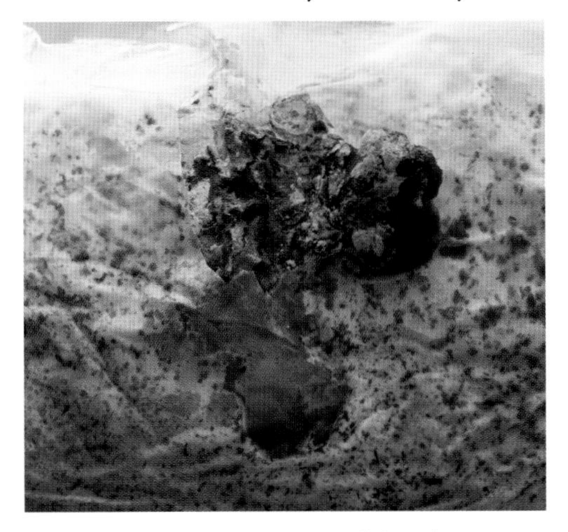

Unfortunately, the optimal conditions for cultivating mushrooms are also ideal for growing mold and bacteria. Cleanliness is paramount.

Molds

Molds can appear in a broad array of colors — black, green, yellow, gray, pink — in your jars. There are many, many types of molds that you may encounter in mushroom cultivation. Here are some of the most common ones.

Blue-Green Mold (*Penicillium* spp.)

There are hundreds of species in this genus, and most are impossible to tell apart without the use of a microscope. The blue-green molds that appear in your projects could be the same species used to create the antibiotic penicillin. Under a microscope, identification of the genus is fairly easy because of the long chains of conidia (spores) that form on a branched base, but as with many types of mold, identification of the actual species can be a chore, even for someone trained in the field.

This mold is extremely common for the home mushroom grower, so the main thing you need to know is that it often appears as a green or bluish-green colony. This colony is usually not exceptionally fast growing, but if you notice it in any jars, they should be discarded immediately. I have seen mushroom mycelium completely envelop a colony of blue-green mold, but that does not mean the colony is dead. If you were to break up the jar and transfer it, you would be spreading the contaminant all around the next phase of your project, and it would surely fail. This mold is most commonly found in jars of colonizing grain. As with most contaminants in grain jars, there is little hope of saving the substrate. It should be thrown out immediately.

Green Mold (*Trichoderma* spp.)

Trichoderma (*trike*-oh-derma) is one of the fastest-growing molds you're likely to encounter. It's most commonly associated with casing layers, but it can also affect jars. I've also seen it appearing with some regularity in straw projects. In nature, there are about 30 species of this mold, most commonly found in the soil and on dead wood.

Trichoderma often appears as mycelium when it's young. At this phase, it's a bright white mass of mold, but as it ages, it turns a bold green. Green means that the colony has begun to form and release its spores, so it should be removed immediately. With a bit of experience, you should be able to tell the difference between the white of a young trichoderma colony and the white of mushroom mycelium. In general, mushroom mycelium is wispier and not as dense, especially while it's still colonizing. Trichoderma often appears as a solid white colony, without clearly identifiable rhizomorphs or mycelial threads. Trichoderma is the primary mold associated with the destruction of agaricus crops on commercial mushroom farms. Many of these farms use fungicides to prevent this mold from appearing in the peat moss of the casing layers.

There aren't really any effective ways to combat this mold once it has begun to sporulate, especially if it establishes itself in a colonizing jar or bag. Remove the contaminated project from the rest as soon as it's spotted. If you find green mold in a casing layer, it may be possible to remove the infected portion and save the rest of the casing. This process is detailed on page 141.

Cobweb Mold (*Hypomyces* spp., *Cladobotryum dendroides* or *Dactylium dendroides*)

One of the most common contaminants for mushroom cultivators, cobweb mold is most often found growing on cakes or casing layers. It's virtually identical to mushroom mycelium, which makes it very difficult to detect. There are some features you can learn to look for that may help you spot it early. First, cobweb mold is somewhat grayer than the pure white of mushroom mycelium. Second, it has a wispier growth habit than mushroom mycelium. Finally, it tends to grow much faster than mushroom mycelium. It can grow from a small spot on the casing layer to a softball-sized infection in just a couple of days.

If this mold grows near developing mushrooms, it can colonize and kill them as well. If you identify it early, you can use hydrogen peroxide to combat it (see page 141).

Aspergillus spp.

While most molds you encounter in mushroom cultivation are not particularly pathogenic to humans, aspergillus is an exception. If you inhale aspergillus spores in large quantities, you may contract a disease known as aspergillosis. When confined to the lungs, symptoms can include cough, difficulty breathing, and chest pains. If the disease progresses into the body, it can cause problems involving the lungs and kidneys.

Aspergillus species can form many different colors in a colony, including gray-green, yellow-green, or black. Because of the broad range of morphology and the difficulty of properly identifying the species, take great care when working with and disposing of any mold you encounter, and avoid breathing in large quantities of any spores.

Lipstick Mold (*Geotrichum* spp.)

As the common name suggests, this mold gets its name from the pinkish color it acquires as it ages. The only time I ever encountered this mold was in PF jars when I first started growing. While not particularly common anywhere but in PF jars, it's an easy contaminant for the beginner to identify. If you encounter this mold in your jars, you need to reexamine your sterile procedure.

Bacterial Contaminants

Bacteria can also be a significant problem in mushroom cultivation. Bacteria often appear as a clear or grayish slime, usually accompanied by a sweet or pungent smell. Here are two of the most common bacteria in mushroom cultivation.

Wet Spot (*Bacillus* spp.)

Most commonly present on grain spawn, this bacterium appears as a clear, grayish slime on the grain in the jar and produces a pungent odor. Mushroom mycelium will not colonize areas of the grain that are infected with bacteria. Bacterial endospores are a common problem with spawn-makers, as they may not be killed during a normal sterilization cycle. If you're having problems with bacteria, consider using the soaking method for your grain, increase your sterilization times, or both. For more on endospores, see page 201.

Bacterial Blotch (*Pseudomonas* spp.)

This is one of the few diseases that affects mushrooms after they have started to form. Sometimes the mushroom caps will appear to have spots and are decomposing before the caps themselves are fully formed. This is caused by a bacterial infection. The spots may first appear slimy, and they eventually form foul-smelling lesions on the mushroom. If you encounter this disease on mushrooms that are forming, remove the infected mushrooms immediately. You should also decrease the humidity in your growing environment and increase the amount of incoming fresh air.

Contamination is a fact of life for the home cultivator or commercial grower. It's best to remain optimistic, but take the time to think through all your sterilization

Contamination: Inside and Out

When I first began to grow mushrooms, everything I read told me to be thoroughly "contamophobic." Almost everywhere I looked I saw a possible source of contamination, so I cleaned, sanitized, and sterilized nearly to the point of obsession. I think that helped improve my success rates early on, because I had only rare battles with contamination. As I gained experience, I began to recognize some shortcuts and to narrow down the most likely sources of contamination. But when working with agar, for example, I would still shower, put on clean clothes, wear rubber gloves, pressure cook all my utensils to sterilize them, work in front of a flow hood, use Parafilm (a flexible paraffin film) around the plates, and have a bottle of alcohol next to me at all times. Following all these steps was the only way I could have peace of mind. This is how I grew up as a grower.

Fast-forward several years to my first actual mycology lab during graduate school. On the first day of lab, we performed a procedure, quite common with mold remediators, known as *settle plates*. Often used to measure the spore load in homes, settle plates will give you a general idea of the mold levels present in a given area. Our instructor had several plastic bins of petri dishes filled with agar (these are the settle plates), and each of us was told to grab four petri dishes. We uncovered the dishes in open air, and left two on the lab bench indoors and two on the ground outdoors for 30 minutes. When the time was up, we covered them back up, let them incubate for 2 weeks, and waited to see how many growth colonies appeared.

I'd been growing mushrooms for about 5 years at that time, so I guessed there would be significant growth in all the dishes, as they'd been left open for 30 minutes. To my surprise, only two or three colonies grew in most of the indoor plates, while those in the outdoor plates were all TNTC (Too Numerous to Count). This clearly demonstrated what our instructor had told us: the air outdoors contains about 10 times the number of fungal spores as the air indoors; thus, you can safely open petri dishes in the lab for transfers without the use of a cleanroom or laminar flow hood. If only two or three colonies grow after leaving the plate open for 30 minutes, you can reasonably expect to have no contamination when opening the plate for the several seconds it takes to do an agar transfer.

Still I advise against doing lab work in the open air, even if the contaminant levels in your home are low. My advice is to take as many steps as possible to reduce the potential for contaminants. Even though I now have a more realistic perception about contaminants in the environment, I still think like a contamophobe, and suggest that you do the same, because it will greatly increase your chances for success.

procedures and try to determine how the contamination occurred. Every grower learns by trial and error, and the errors are often more instructive than the successes, as they can help you prevent mistakes on future, larger batches of mushrooms.

Disposing of Contaminants

Because some molds can be harmful to your health, and most can be detrimental to your cultivation projects, never open contaminated jars indoors. Opening jars indoors only spreads the mold spores around your house, unleashing the potential for future contamination. After you open the jars, dispose of the substrate outdoors. Then change your clothes before going anywhere near your lab and growing area. Mold spores stuck to your clothes can be tracked back inside with you.

If a jar has a serious amount of contamination, and you want to save the jar, consider pressure cooking the jar before you open it. This will kill all the mold spores inside and prevent them from spreading. Opening a jar that has a lot of contamination will often result in a visible cloud of spores being released. If the whole jar is green (or some other color) and you don't want to pressure cook it before opening it, just throw the entire jar out. Saving one jar is not worth the potential for contaminants long into the future.

After you have dumped out the contaminated jars, wash them out with a hose outdoors before you bring them back indoors. Once the jars are back inside, wash them out with soap and water.

Using a Glovebox

Keeping your home as clean as possible is only a small part of the race against molds and bacteria. Although the air indoors may be cleaner than the air outdoors, it's not nearly clean enough for many of the procedures used in mushroom cultivation. To perform them successfully, you'll need some equipment specifically made to produce environments clean enough for sterile work: a glovebox or a flow hood.

A glovebox is a sealed or semi-sealed box that creates a sterile, still-air environment around the materials with which you're working. Spores and other contaminants settle out of the air to the bottom of the box, leaving you with sterile air to work with. While not a perfect solution, a glovebox is a great low-cost option that will increase your chances for success.

Gloveboxes are commonly used by beginners for procedures such as making syringes and liquid cultures, inoculating jars, and grain transfers. At a minimum, I would recommend the creation of a simple glovebox prior to your first grow. It will get you into the habit of adopting sterile procedures and will ultimately be beneficial to your efforts. The next best method is the construction of a flow hood, but that requires a serious investment of time and money, and I don't recommend it until you have several grows under your belt. (For information on the construction and use of flow hoods, see chapter 5.)

Many variations of the glovebox can be made, depending on your individual needs. One variation that I would *not*

Glovebox Gunshot

I've made several blunders over my years of mushroom cultivation. One of the most dramatic, which occurred when I first started growing mushrooms, involved a glovebox and a can of Lysol.

Very early in the growing process, I realized that I'd need a sterile work environment to maximize my chances of success. After looking online at many options, I settled on a design very similar to the glovebox shown on the following pages. Most of what I was reading at the time suggested Lysol was one of the best materials to use to sterilize everything in the box. I'd also read that utensils such as scalpels and needles must be flame sterilized before use. I knew that Lysol was flammable, but I thought that if I let it sit long enough, the fumes in the box would dissipate. So, prepping for the first inoculation, I loaded my jars and everything else I needed into the glovebox. After an hour or so, I ventured back into my closet workroom, put my hands into the kitchen gloves, picked up the lighter to flame sterilize the scalpel, and gave it a flick.

As you might have guessed, there was a boom, like a gunshot, and I saw a ball of flames. The plastic tub's side latches kept the blast from blowing off the lid and destroying my face. Instead, the lip of the lid directed the flames downward, across my forearms. I threw my body backward, literally through the closet door, to avoid the mini-explosion. My girlfriend came running into the room. Fortunately for me, only the door frame sustained lasting damage. My arm hair was gone, and my forearms were red for several days with a light burn, but it healed quickly.

There are many lessons to learn from this story — *always do adequate research before proceeding, don't use flammable products and a lighter in a sealed environment, think before you act* — but the one I like best is *trust your instincts*. If something doesn't feel right, don't freeze mentally and ignore the potential danger. I remember how timidly I flicked the lighter that day, but it doesn't matter how timid it was, because I still flicked the lighter. When in doubt, *don't*.

PF jars are loaded into a glovebox for inoculation.

recommend, however, is the addition of a HEPA filter unit to your box. Some people cut a hole into the side of the glovebox and attach a HEPA filter/fan unit, commonly available at hardware stores. There are two main reasons why I believe this is a bad idea. First, it defeats the primary goal of a glovebox — to create a still-air environment. Constant airflow into your glovebox will create swirls of contaminant particles in the box and around your work. When you open jars, these swirling contaminants will be much more likely to get in. Second, while HEPA filters are designed to clean air with 99.98 percent efficiency, the units' outlet ports are not adequately sealed. They often allow the introduction of contaminated air into the outlet stream after the filter. I have not encountered any readily available air filter units with filter housing seals that are suitable for glovebox work. Please keep this in mind when you compare glovebox designs.

Using a glovebox to inoculate PF jars helps prevent contamination from mold spores and bacteria.

How to Make a Glovebox

Materials

- Heavy-duty plastic storage bin
- 18" × 12" (46 × 30 cm) sheet of clear acrylic
- Silicone sealant
- Strong glue
- Knife
- Loose-fitting rubber kitchen gloves
- Wire coat hangers
- 10:1 water-to-bleach solution in a spray bottle; or 70–90% isopropyl alcohol
- Paper towels

1 **Make a viewing port.** Most heavy-duty plastic bin lids are not clear. If yours is not, you'll need to start by making a viewing port in the lid. Mark off an area of your lid 2" (5 cm) smaller on all sides than the size of your acrylic sheet. Cut out the outlined hole in the lid, being careful not to crack the plastic as you proceed. A heated knife will cut through the plastic more easily. Place a layer of silicone sealant on the lid around the opening and press your acrylic sheet into the sealant. Let dry.

1

2 **Cut the armholes.** Cut two 5" (13 cm) circles into one side of your bin. Position them slightly farther apart than the width of your arms at your sides, or at a comfortable working distance.

2

3 **Make the glove supports.** Fashion two 6" (15 cm) circles out of the wire coat hangers. Insert one kitchen glove through the wire circle, leaving 2" (5 cm) of the glove free to fold over the wire. Fold the end back up around the wire circle and use strong glue to secure it. Repeat the same procedure with the other glove.

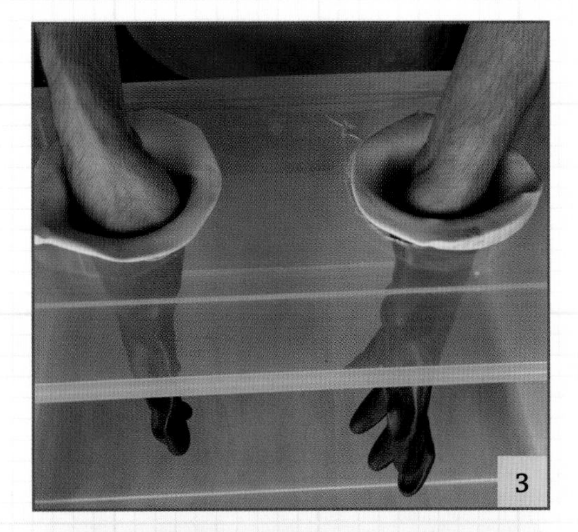
3

4 **Seal the gloves into the armholes.** Put the gloves into the holes in the sides of the bin. At the wire area, seal each glove to the plastic hole opening using glue or silicone.

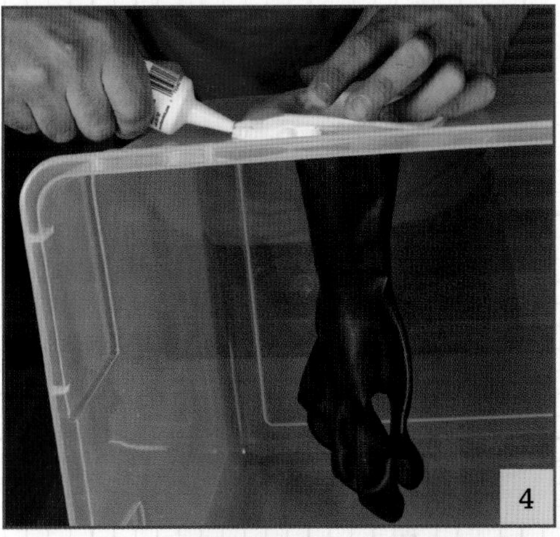
4

5 **Start using your glovebox.** Open the lid and place all the items you'll need for your sterile procedure into the box. Seal the lid and allow your box to sit for an hour or so. This will allow contaminant particles in the box to settle before you begin your procedure. A common disinfectant to use within a glovebox is a 10:1 water-to-bleach solution. Isopropyl alcohol or Lysol can be used as disinfectants in a glovebox, but only if you do not need a flame for your sterile work. Using a flame in the presence of isopropyl alcohol or Lysol can cause a fire or explosion.

5

Create an Inoculation Room

Once you've made a glovebox, you'll need a secluded place to use it. The ideal place for an in-home inoculation room is probably a closet. I set up a table in a bedroom closet for the sterile work of my first grows. Closets have several advantages over other parts of the house. Nearly still-air environments, closets are generally isolated from the air-circulation systems in the rest of the home, and they usually have doors you can close to avoid extraneous air currents. If your closet is carpeted, it's best to lay some plastic down over it to limit the amount of dust you kick up whenever you enter. Also be sure to remove any clothes, towels, and other similar items from the closet. Clothes tend to harbor contaminants, which are released whenever the clothes are moved or jostled.

It's also possible to inoculate jars in a clean bathroom without the use of a glovebox, if your home has a relatively low spore count. Bathrooms usually have tile floors rather than carpeting, so the amount of dust in bathroom air is usually lower. Also, bathrooms usually consist of hard, smooth, durable surfaces that are easy to clean with heavy-duty cleaning products. The main drawback of bathrooms is their humidity, which encourages molds to proliferate. I prefer closets to bathrooms for inoculations and other cleanroom work, with or without a glovebox. But if a closet is not available, the bathroom is the next best option.

Whichever room you choose to use, clean all surfaces in the room with a solution of 10:1 water to bleach, isopropyl alcohol, Lysol, hydrogen peroxide, or another suitable cleaning product. Many people also choose to spray air sanitizers in the room.

CHAPTER 4

YOUR FIRST GROW

If you want to grow your own mushrooms without the use of a premade kit, this is the place to start. This chapter focuses on one basic method, by far the simplest and most efficient for amateurs learning to grow mushrooms. Originally invented by psilocybe cultivators, the PF Tek method works well for Shiitake and Oyster mushrooms, two species that are good choices for beginners.

The PF Tek Method

This method, originally published in 1995 by an individual who used the pseudonym Psilocybe Fanaticus (PF), involves steaming vermiculite and brown rice flour in small, half-pint canning jars. Once cool, the concoction is injected with a syringe of spores or a liquid culture. Growers ultimately end up with colonized "cakes" of mycelium that are fruited in a small terrarium. PF was arrested in 2005 and prosecuted by the Drug Enforcement Agency for growing illegal hallucinogenic varieties, but he lives on through his legacy — this simple method for home mushroom cultivation.

The materials required for preparing your first PF jars are relatively basic. Brown rice flour and vermiculite will form the substrate.

There are several key reasons why this is a good method for beginners. First, it uses materials that are available everywhere. Vermiculite is available at most garden centers; brown rice flour is available at health food stores and many groceries. Second, it is possible to sterilize this substrate by simply steaming the jars in a pot on the stove. Most of the other methods described in this book require a costly pressure cooker. Third, this method uses spore or liquid culture syringes, commonly available online, as a starting point. Finally, this method only requires a fruiting chamber the size of a small plastic bin. If you're a beginner, I encourage you not to prepare more than 24 PF jars until you've fruited at least one smaller round of jars successfully. Trying to overproduce before you understand all the methods only increases your chances of failure. Start with something small and manageable, and you will be rewarded.

Using the PF Tek Method

Materials

- Half-pint wide-mouth canning jars
- Vermiculite (or sawdust, for wood-loving species)
- Brown rice flour
- Water
- Measuring cups
- Mixing bowls
- Paper towels
- Alcohol & swabs or cotton balls
- ⅛" (3 mm) nail
- Hammer
- Masking tape
- Aluminum foil
- Large pot with lid or pressure cooker
- Spore or liquid culture syringe
- Bleach solution
- Lighter

PF Tek Fruiting Formula (per jar)

- ½ cup (8 g) vermiculite or sawdust
- ¼ cup (60 mL) water
- ¼ cup (45–50 g) brown rice flour

Time Required

Prep: 30 minutes–1 hour

Sterilization: 1½ hours

Inoculation: 20 minutes

Incubation: 2 weeks

Dunking: 24 hours

Fruiting: 2 weeks

Substrate Preparation for 1 Dozen Jars

1 **Prepare the jars.** Hammer four holes through each lid, using a ⅛" (3 mm) nail. Wipe the interior of each jar with alcohol.

2 **Mix the substrate.** In a large mixing bowl, combine 6 cups (96 g) of vermiculite or sawdust and 3 cups (.75 L) of water. Mix thoroughly. Add 3 cups (500–600 g) of brown rice flour to the vermiculite mixture and mix thoroughly. You should now have enough substrate to fill about 12 jars.

3

5a

4

5b

3 **Fill the jars.** Fill each jar with the vermiculite mixture to the level of the lowest ring band. Do not pack it down. Wipe the tops of the jars clean with a paper towel. Fill each jar to the top with a layer of dry vermiculite. This serves as a contamination barrier.

4 **Place the lids.** Place the lid and screw on the ring band. Tape over the holes using masking tape. Fold the tape over to make a small tab — this makes it easy to pull off after it is sterilized.

5 **Sterilize the jars.** Cover each jar with foil. I usually make 8" × 8" (20 cm × 20 cm) squares of foil, then press them on. To steam the jars, put 2" (5 cm) of water in a large pot. Add a small rack or other raised surface inside the pot to keep the jars out of the water, and place the jars on the rack. Put on the cover and steam the jars for 60–90 minutes. If using a pressure cooker, leave the jars in for 45 minutes at 15 pounds per square inch (psi). Let the jars cool in the pot. Don't open it until you're ready to inoculate them. Allow your jars to cool completely to room temperature before inoculation.

6

7

6 **Inoculate the substrate.** Move the jars into your glovebox or in front of your flow hood. Remove the foil from the jars. Put your syringes, bleach solution, paper towels, alcohol swabs, and lighter into your glovebox.

Spray a piece of paper towel with the bleach solution and use it to wipe down the sides and bottom of the box, the syringes, and any other tools.

Shake the syringe to distribute the spores or culture evenly within the fluid. Remove the plastic needle cover and flame sterilize the needle. If your syringe arrives disassembled, you may need to attach the needle before you can sterilize it. Sterilize by holding a flame to the needle until you can see that it is red hot. Let the needle cool, or fast-cool it by wiping it down with an alcohol swab.

Pull back the tape and inject about ¼ cc of fluid from the syringe into each hole of each jar lid. Cover the hole again after each injection. Repeat this process for all jars.

7 **Incubate.** Place your inoculated jars in a warm, dry, dark place for colonization, or make an incubator for this process (see the next section). For most species, you'll want the temperature of the incubation area to be in the 75–85°F (24–29°C) range.

Let the jars sit and incubate for 2 weeks. You should see initial growth within your jars in 3 to 5 days, although it can take up to 7 days or more. Full colonization of the jars will usually take 2 to 3 weeks. See page 63 for more information on incubation.

If you see growth in any of the jars that is any color other than white, remove those jars and dispose of them immediately. Common colors are green, pink, or black. Never open contaminated jars indoors. See page 45 for more information on contaminants.

8

8 **Dunking.** Once your jars are fully colonized and ready to be placed in the fruiting chamber, you should dunk them in water for 24 hours before proceeding. Dunking gives the jars a burst of extra moisture before the fruiting process begins, and results in better yields. Working over a sink, simply unscrew the lid of each jar, fill the jar with water, and screw the lid on again. The colonized cake will probably float, so when you put the cap back on, you may need to force it back down into the water, causing some of the water to overflow. Store the jars in the refrigerator for this 24-hour dunking period. (See the dunking section in FAQ for PF Tek, on page 61, for more information.)

9 **Fruiting.** Place the jars into a fruiting chamber. (See page 70 for information on constructing a fruiting chamber.)

You may want to consider double-end casing your cakes (see page 67) to improve water retention for the fruiting process.

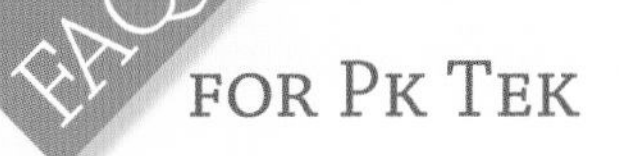

Substrate Preparation

Do I have to use wide-mouth jars?

Yes. You'll ultimately want to remove the colonized cake from the jar in one piece. This is possible only with wide-mouth jars, which do not taper inward beneath the lid. If the jar tapers, the cake will not slide out in one piece.

What type of vermiculite should I get?

Vermiculite is typically sold in a range of consistencies from fine to coarse. Fine vermiculite is almost like sand. The coarse vermiculite contains larger particles that are better for mushroom cultivation. Coarse is ideal.

Can I use tap water for this process?

In most municipalities, tap water will work just fine. If you have any concerns about your water quality, use store-bought spring water.

I can't find brown rice flour. Will other types of grains work?

Yes, other types of grains will work with this process. Rye flour is one alternative.

Can I grind my own rice?

Yes, some people choose to grind their own rice to allow better control over the particle size of the flour. I would not bother, although if you already have a coffee grinder or mill, it may be cheaper to buy whole grain. Some people say freshly ground works better, but I am disinclined to believe it's of substantial benefit, if any at all.

My mixture gets clumpy. Why does this happen?

The most likely cause is that the flour is clumping up. Be careful to add the flour only after the vermiculite and water are fully mixed in the bowl. Adding the ingredients in this order should prevent clumping and yield a more uniform mix.

Sterilization

Should I buy a pressure cooker?

The short answer is yes. Sterilizing your jars in a pressure cooker will always be better and lead to more successes than boiling the jars. I would suggest getting one as soon as possible, even when you're just starting out. Boiling works merely okay as a way to get started without much investment.

What is the purpose of the foil over the jars?

The foil helps keep extra water out of the jar while it is being sterilized. It also helps keep the injection sites free from contaminants until you are ready to inoculate the jars.

Is it bad to get water in the jars?

Yes. This is why the jars are raised up out of the water while they are being boiled. If the jars fill up with water during the sterilization process, they need to be remade. Mycelium will not grow in an environment that is full of water. The easiest way to tell if water has penetrated the jars is if they are significantly heavier after sterilization.

Should I sterilize the jars for 60 minutes or 90 minutes?

If you are boiling, and doing it right, you should have very few issues with contaminants at 60 minutes. If you are having issues, try upping your boiling time to 90 minutes. Or you can just start at 90 minutes from the beginning and never look back. The only reason to do 60 is because it works and obviously takes less time. If you're using a pressure cooker, 45 minutes at 15 psi will be sufficient.

How high a boil should I use?

A low boil will work fine. It doesn't have to be a violent, rapid boil. If you have steam coming out from the side of your pot lid, the boil is high enough.

Can I overcook my substrate?

Probably the only way this is possible is if your cooker/steamer runs dry. If this happens, your substrate may visibly burn. Otherwise, it's not possible to overcook your substrate using the timeframes and pressures given.

Inoculation

Is a glovebox required for inoculation?

No, the jars can be successfully inoculated in open air. A glovebox is helpful for beginners, however, as it reduces the potential for contamination.

Would adding more than ¼ cc per inoculation site speed things up?

It will most likely speed it up a very small amount, but not enough to offset the additional number of jars you could otherwise have produced with that extra inoculant.

Where should I inoculate in the jar?

Most people choose to inject the substrate near the glass of the jar. Inoculating near the glass, as opposed to the center of the jar, allows you to keep track of early germination and to see if there is any contamination present at the inoculation site. It also allows you to see how much fluid you have injected.

Incubation

My cakes started colonizing but then stalled before 100% colonization. Why?

One of the most common reasons for this is a lack of fresh air. Were the lids turned upside down from the start? If so, try removing any tape that is covering the holes on the jar lids and loosening the ring band. This should allow for more fresh air to enter.

At what temperature should I keep my cakes?

I have yet to come across a species that will not colonize quickly at around 75°F (24°C). Many species will grow slightly quicker at up to 85°F (29°C). For this reason, some people choose to build incubators.

Dunking

When exactly should I dunk the cakes?

I would wait for 3 days or so after your cakes achieve 100 percent colonization. This will give internal parts of the cake a chance to colonize. With Shiitake, I would wait for an additional week before opening them for fruiting. You'll also want to cold shock them in the refrigerator while they're being dunked.

Incubation and Incubators

Many mushroom species have an optimum temperature for their vegetative growth cycle that is slightly higher than 70–75°F (21–24°C). Thus, if your goal is fast growth, your colonizing jars should be stored in a place that has this ideal temperature. Any place in your home will work if it maintains a slightly higher temperature than the average in the rest of your house and is warm, dry, clean, and free of mold or other potential sources of contamination.

If you can't find a place in your home with a consistent ideal temperature, you can make an incubator. Shortly after I started growing mushrooms, I made an incubator similar to the one described here and used it for a few months. Since then, I have simply let my jars colonize on a shelf in the lab room at room temperature, so building an incubator is entirely optional. If your gut tells you this is something you need, or if you'd like to experiment, go ahead. Incubators are not terribly expensive and will most likely save you a day or two in colonization time. It's up to you whether that day or two is worth it.

Optimum Vegetative Temperature Ranges of Common Species	
Shiitake	70–80°F (21–27°C)*
Oyster	70–85°F (21–29°C)*
Maitake	70–75°F (21–24°C)
Lion's Mane	70–75°F (21–24°C)
Reishi	70–80°F (21–27°C)
Psilocybe cubensis	80–85°F (27–29°C)

* substrain dependent

Make an Incubator

Materials

- 2 heavy-duty plastic bins
- Fully submersible fish-tank heater
- Zip ties
- Water
- Towel or cloth

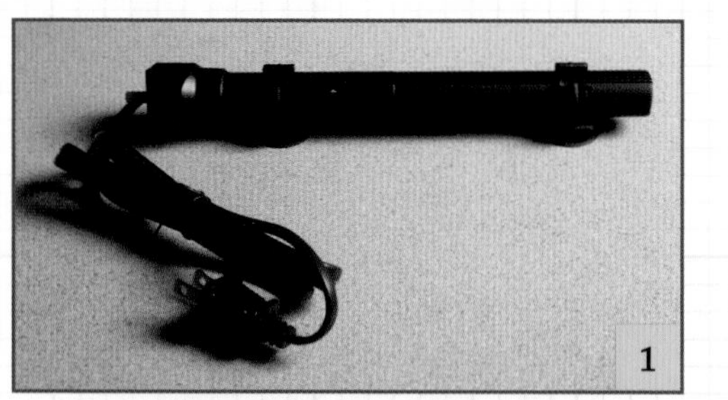
1

2

3

1 **Know your tools.** Fish-tank heaters are a potential fire hazard, so read all directions before using one. The fish-tank heater should come with suction cups that allow it to attach to a surface.

2 **Set up the heater.** Attach the heater to the bottom of one of your bins. Add water to fill the bin several inches above the top of the fish-tank heater. Turn on the heater, and allow it to warm the water to a few degrees above your desired colonization temperature. Once your second bin is attached, you'll no longer be able to make temperature adjustments without removing it.

3 **Place the second bin into the first bin.** Force it down so the water level rises up the sides in between the two bins. Do not let the second bin touch the heater. Zip-tie the two bins together. Add water until the water reaches at least 6" (15 cm) up the side of the bins. Cover the bin opening with a towel or other cloth.

An Alternate Incubator

This is the perfect option if you want to store your incubating jars. You can store them in a file cabinet, cardboard box, entertainment center, speaker, cooler, etc. Simply open a hole the size of your heater in the lid of a quart jar, fill the jar with water, and insert a 20- to 50-watt fish-tank heater into the jar.

Contamination in Jars

When mushroom cultivation fails, it's usually because of contaminated substrates. Contamination is most likely to occur during the incubation period, and it's a matter that you'll need to take seriously and deal with appropriately.

Contaminants to mushroom cultures generally present themselves as molds of a color other than white. Greens, reds, grays, blacks, and yellows are all common mold colors in mushroom cultures.

This jar of Shiitake culture is contaminated with mold and must be disposed of. To prevent spreading mold spores, never open contaminated jars indoors.

If you notice any of these colors while a jar is colonizing, do not try to save your substrate. Under these circumstances, you have two options. The first is to throw the entire jar away without opening it. If you open a jar that has large amounts of colored spores growing inside, you'll almost certainly spread the spores across yourself and your environment. If you open the jar indoors, you'll make future grows more difficult by introducing many more contaminants into your home. Never open a jar that has a large amount of mold growing inside. Throw the entire thing away.

As you gain experience, you'll be able to spot potential contaminants earlier in the colonizing process. That brings us to a second option. If there's only a small amount of contamination inside the jar, it may be possible to save the jar. You can never save the substrate, just the jar that contains it. So if your spot of contamination is small, take the jar outdoors, open it, and throw away the substrate. After you open and empty the jar, wash it out thoroughly with a hose. If a large puff of spores (looks like smoke) comes out of the jar and blows across your body, remove any of your affected clothes before going back indoors. Otherwise, you'll risk tracking large amounts of potential contaminants back in with you.

Dunking

Dunking is the process of submerging a cake or other substrate in water for a period of time (usually 24 hours) after full colonization and before it is placed into the fruiting chamber. This gives the substrate a sudden influx of additional moisture that will encourage better yields from the cakes you have prepared. As your cake was incubating for several weeks, much of the moisture was absorbed into the growing mass of mycelium. Some of the original moisture was also lost to evaporation. As the mycelium begins to form mushrooms (which are 90 percent water), a fully hydrated substrate will help provide a sufficient reservoir for their growth.

The process for dunking cakes is very simple, but be sure that the entire cake is fully colonized before you begin. Fully colonized cakes look white throughout. Most cultivators will allow seemingly colonized cakes to remain in the jar for several additional days before proceeding to ensure full colonization.

PF jars can become contaminated with mold and must be discarded.

If you open a cake and find that part of it is not colonized, just wash off the uncolonized portion under running water. Don't leave uncolonized portions on the cake for the fruiting process — they are likely to grow mold.

When you are certain that your cake is fully colonized (it will be white throughout) and ready to be placed in your fruiting chamber, take your jar to a sink. You may use normal tap water for dunking the cakes. Open the jar and allow the water to fill it through the vermiculite layer on the top. When your cake begins to float, you have enough water. Screw the lid back on, forcing the cake slightly back down into the water. This will probably force some water to flow out over the sides of the jar and into the sink.

Sometimes the cake will not float, because the mycelial mass has attached itself to the side of the glass jar. This is fine, too. Just fill the jar with water to the brim and screw the lid back on.

After the jars are filled with water, it's best to put them in the refrigerator during the dunking period. For most mushroom species, a temperature drop is a signal to their culture that it's time to begin the fruiting process. Also, low temperatures help slow down the growth of any contaminants. If any were introduced when you opened the jar, they will not mature quickly under cool conditions.

Double-End Casing

Before you place your dunked cakes in a fruiting chamber, you should consider one more procedure called double-end casing. A casing is a layer of nutrient-free material that is placed on a fruiting substrate to help the substrate retain moisture. Commercial cultivators often use peat moss for this layer, but vermiculite is the best choice for home growers using the cake method. Just as dunking helps hydrate the substrate before fruiting, casing your cakes adds an extra layer of material on both the top and bottom that will help retain moisture throughout the fruiting process. The casing process can be performed in your usual work space; gloves are optional.

After a colonized cake has been dunked, it's helpful to create a "double-end casing" with damp vermiculite to help the substrate in the cake retain moisture. The cake is then placed in the fruiting chamber.

Sterilizing Vermiculite

Vermiculite taken directly from the bag will usually be sterile enough to use for a casing layer. If the bag is old or has been sitting out for a while, you can ensure that you have sterile vermiculite by microwaving it for 5 minutes.

Don't use ordinary plastic food containers to microwave vermiculite, because they will quickly melt. Vermiculite in the microwave gets very, very hot and will melt all but the most durable plastics. Instead, use a glass bowl. I also recommend microwaving your vermiculite when it's dry. If you microwave it when it's wet, you'll have to wait a very long time for it to cool down. Adding water to the vermiculite after you microwave will cool it down and ready it for more immediate use.

Double-End Casing Method

Materials
- Colonized cakes
- Several cups of vermiculite
- Water
- Spoon

1 **Create the base.** Sterilize your vermiculite if necessary (see page 67) before you begin. Place vermiculite in a small bowl. Moisten the vermiculite to field capacity. (The vermiculite should not drip water unless you give it a light squeeze.) Remove the lid of your colonized jar and set it on the counter. This will serve as the base of your mushroom cakes in the fruiting chamber.

Spoon enough sterile, moistened vermiculite from the bowl to fill the jar lid.

2

3

2 **Remove the cake from the jar.** If the cake is sticking to the sides of the jar, give it a light tap on the counter to break it free. Be careful not to damage or otherwise break apart the cake itself. Remove the vermiculite layer from your colonized cake in the jar and discard it. You only want to remove the vermiculite layer. If the mycelium has grown into the vermiculite layer, it's okay to remove that mycelium from the cake.

3 **Set the cake on the fresh layer of vermiculite.** Spoon more vermiculite from the bowl onto the top of the cake.

4 **Stack the cakes.** If you're stacking your cakes, place another colonized cake on this layer, and top it off with a final layer of vermiculite.

4

Fruiting Your Cakes

After your fully colonized PF cakes have been dunked and double-end cased, they're ready to be placed in a *fruiting chamber*, also known as a *terrarium*. A fruiting chamber will give your mushrooms the best possible conditions for fruiting correctly.

Small Fruiting Chambers

There are many types of fruiting chambers out there, and some are fancier than others. I would strongly encourage you to shy away from fancy, elaborate chambers, as they won't work any better than the simpler ones. Until you understand the fundamentals of mushroom cultivation and how its many variables affect each other, I recommend that you keep it simple. Don't introduce unnecessary variables; you'll only have to work through all of them to diagnose contamination problems if you have an issue. Simpler setups always have less chance of failure, are easier to manage, and will make your life more efficient and more pleasurable.

Keep in mind that the suggestions in this chapter are for small-scale fruiting PF terrariums, and are not ideal for larger-scale setups. Suggestions for larger setups are discussed in later chapters.

This fruiting chamber is set up and ready to receive cakes.

There are four main variables to consider when planning a fruiting chamber: size, humidity, fresh air, and light.

Size

Your fruiting chamber must be large enough to hold the quantity of jars you have produced. PF jars will fruit when stacked one on top of the other, so if you have prepared twelve jars, the bottom surface area of your fruiting chamber must be larger than six jar lids spaced several inches apart. The height of the chamber must be a minimum of three canning jars high above the perlite layer.

The next consideration related to size is CO_2. CO_2 is produced when the mushroom mycelium eats the substrate, and this gas will build up in your fruiting chamber. The smaller the chamber, the faster the CO_2 will build up and begin to inhibit growth. Thus, in a smaller chamber, you'll need to fan more often.

Humidity

Mushrooms like moist places, and in the wild they fruit after rain comes. Thus, it should be no surprise that humidity will be one of your main concerns in the fruiting chamber. There are several key methods for maintaining humidity on a small scale. The primary one is perlite.

Perlite

Perlite is a white, rocky material mined from the earth. The interiors of perlite rocks are very porous, so when they're moistened, they absorb water and release it slowly through evaporation. This makes perlite a perfect material for maintaining the proper humidity in a small fruiting chamber.

The most common way to use perlite is to soak it in a bowl of water, and put a 2- to 3-inch (5 to 8 cm) layer of it in the bottom of your fruiting chamber. This should maintain a decent humidity level in the chamber for several weeks before it needs to be changed. If your chamber is a little larger, consider a layer up to 4 inches (10 cm) deep.

One of the best things about using perlite in the bottom of your fruiting chamber is that it does not contain very much nutrition. That makes it hard for molds and other types of contamination to get a foothold and ruin your attempts at cultivation.

Perlite

Air Bubbler

The best addition to perlite in a small fruiting chamber is a fish-tank air bubbler. This consists of a fish-tank pump, plugged into a wall outside the chamber, from which a hose runs to an air bubbler in a glass of water in the bottom of the chamber. The glass of water is filled one-half to three-quarters full. Using the air bubbler in combination with perlite will not only increase the humidity slightly, but will also continuously add fresh air.

Air bubbler

Spray Bottle

Any standard spray bottle that shoots a fine mist rather than a strong jet of water will work fine. Be sure to avoid bottles that formerly contained chemicals or other toxic substances.

It's best to air out your fruiting chamber two or three times a day. Whenever you air out the chamber, give it three or four mists with your spray bottle as well; however, don't mist your cakes or fruiting mushrooms directly. If the walls of your fruiting chamber are dry, give them a couple of sprays to keep them moist. It's also fine to mist over your cakes gently from well above them, but don't blast them with a strong jet of water. If you spray from above, let the atomized water droplets fall gently to the surface of your cakes.

Humidity Levels

Use the spray bottle on an as-needed basis, in combination with the perlite and the air bubbler, to attain the proper humidity in a small terrarium. For PF cakes, keep the humidity in your fruiting chamber above 90 percent. At this point, humidity gauges or other monitoring are not necessary. If you have several inches of wet perlite in the bottom and an air bubbler, and if you're performing hand misting, your levels should stay near the right point without any other monitoring or tools. If you really want more data, a simple humidity gauge, available at most garden centers or big box stores, will serve that purpose.

Fresh Air

A third important consideration when designing your fruiting chamber is fresh air. As previously mentioned, growing mushroom mycelium produces CO_2, which tends to build up within a closed fruiting chamber. Visible symptoms of high CO_2 levels are skinny, elongated stems and/or small caps on the growing mushrooms. The easiest way to ensure that CO_2 levels do not get too high is to manually fan your chamber two or three times a day. Just use a scrap piece of cardboard or the lid of your bin as a fan to force fresh air down to the level of the cakes. Fanning semi-vigorously for 5 to 10 seconds, two to three times a day, should do the trick. If you have a fairly small fruiting chamber relative to the size of your mycelia mass, I would increase the number of fannings to four or five times a day.

Please remember that these suggestions are not chiseled in stone. If you miss several fannings, it's not the end of the world, and your mushrooms will probably still grow fine. I've given you suggestions for what's best in an ideal world, but since we're only hobby growers, other things need to take priority sometimes. Going through this process and growing out your mushrooms several times will give you a feel for what they need, as well as what they like and don't like, so try to accommodate your mushrooms as much as possible, and they'll surely accommodate your needs from time to time.

So much more could be said at this point regarding fresh air. For example, there are equations for the ideal amount of air input per unit of volume in the growroom, but these are the concerns of more advanced growers. You don't even need to think about them unless you're considering small-scale commercial production.

Light

Contrary to what most people believe about mushroom cultivation, virtually all varieties require light to stimulate initial growth and to complete the fruiting process of forming mushrooms. The only common species whose cultivation often takes place in darkness is the White Button mushroom. These mushrooms were originally grown in caves in France, but most home cultivators don't take the time to grow them.

The best location for your light source is above the fruiting chamber. Most people choose chambers with opaque sides and a clear top. This prevents a lot of light from the sides, and will generally induce your mushrooms to grow upward.

Mushrooms don't need the high intensity lights needed by green plants. Plants get their energy from photosynthesis; mushrooms get theirs by decomposing organic material. Everyday lights of a normal intensity work fine. For example, a 40-watt bulb is more than sufficient for a chamber holding a dozen cakes or less. For slightly larger terrariums, hang a shop light overhead. Depending on the size of your terrarium, you might get by with tap lights. Some people use rope lights; others use only ambient sunlight in the room as the light source. Most mushroom varieties will grow fine with a 12 hours on/12 hours off light cycle, coming from whatever light source you choose.

If you control the four main variables — size, humidity, fresh air, and light — within the proper parameters for the fruiting process, you should have very few problems as you proceed.

Setting Up a Small Fruiting Chamber

Most mushrooms begin to fruit about 2 weeks after they're introduced into the right environment. Some growers use the term "birthing" when they talk about the fruiting process. This simply refers to removing the "child" (cake) from the "womb" (jar). Here is a reliable method for setting up a small fruiting chamber in which you can birth your cakes. Small fruiting chambers like these can also be used to maintain the humidity required for the premade kits and Oyster mushroom toilet paper rolls described in chapter 2.

To set up the chamber, soak your perlite in a bowl of water. Drain the water, remove the perlite, and place a 2–3" (5–8 cm) layer on the bottom of your bin. Place a pint (500 mL) glass of water in your bin and run a fish-tank bubbler stone from the pump into the water. Choose a light source and place it near your fruiting chamber. Place your cakes on a lid or other base and put them into the chamber.

Fruiting without a Chamber

A fruiting chamber offers home growers a more controlled environment for their mushrooms during the fruiting stage, but there are alternatives to using one. The following methods offer less control over individual environmental parameters, but still allow for consistent yields to be attained.

In Vitro Method

The in vitro method is a modification of the PF Tek method that requires no additional fruiting chamber. Use this method if you don't care what your mushrooms ultimately look like — they will be fruited in the closed jar, and compressed between the colonized cake and the glass of the canning jar. This often leads to misshapen, malformed mushrooms. As with the PF Tek method, this method was created by psilocybe cultivators, and it provides for an even easier grow, with fewer steps than the traditional method and an added element of stealth. It will also work for edible species like Shiitake.

To use the in vitro method, repeat steps 1 through 8 as described in the PF Tek method (pages 57–62). Instead of putting your jars in a warm, dry, and *dark* place, introduce them to a warm, dry, and *light* place as soon as they are inoculated. Once the cakes are 100 percent colonized, open each jar and shake off the vermiculite contamination barrier. This will allow fresh air into the jar and provide more room for the mushrooms to grow. Close the cakes back in the jar. These procedures can be done in the open air.

A simple quart jar can be used as a fruiting chamber, if you're only fruiting a few cakes at a time.

For best results, open the jars once or twice a day to allow fresh air into them. Another option is to have the jars upright with the lids sitting loosely over the tops. Open them at least once every few days. Pick the mushrooms when the caps begin to open away from the stem.

After harvesting, fill the jars with cool water and put them in the refrigerator overnight. The next day, empty the water and let them sit again in their jars in artificial light or indirect sunlight. This process initiates the second flush of mushrooms.

Quart Jar Terrarium Method

This method is very similar to the in vitro method but uses a quart jar as the fruiting chamber. This gives your mushrooms the benefit of having space to grow and develop normally. Another advantage is that you don't have to mist the cakes. The main limitation of this method is that it can be somewhat space intensive if you want to grow a large number of cakes. This method works very well for up to 12 cakes at a time.

Contamination During Fruiting

If you've fully colonized your jars without contamination issues, you've made it past the most difficult part of the cultivation process. Freshly inoculated jars are most susceptible to critical issues with contaminants. It's also possible, however, for contamination to occur during the fruiting phase.

One of the main reasons for contamination during colonization is a high CO_2 level (stale air), which encourages mold and bacterial growth. Maintaining fresh air is one of the best ways to prevent contaminants from taking over your cakes as the fruiting phase proceeds. Contaminants may also creep up if the entire cake was not completely colonized before it was introduced into the fruiting chamber. Leaving uncolonized portions of a cake exposed to the open air is like inviting contaminants to a buffet on which to feed and grow. If you open a cake and find that part of it isn't colonized, just wash off the uncolonized portion under running water. Don't leave it on for the fruiting process, as it will most likely grow mold.

Contaminants can also appear in double-end casings. If this occurs, or if contamination appears on the cake itself, remove the infected cakes from your terrarium, wipe down a knife with alcohol, and cut or scrape off the infected areas. Many growers set up a smaller, separate terrarium for cakes that have experienced contamination. This prevents these cakes from infecting others, while still allowing the cakes the chance to fruit.

Rarely, contaminants appear on the perlite in the bottom of the container. If this happens, remove all your cakes, throw away the perlite, and wash out the bin thoroughly. It can be reused, but make sure it's clean before you add fresh perlite. Some growers add a bit of peroxide to the perlite to inhibit mold growth.

FAQ for Fruiting

Is there a spectrum of light that works best? How much light should I provide?
Light in the blue spectrum works best. Most species fruit well with a 12-hour on/12-hour off light cycle. Oyster mushrooms do well in more light than most species, so I would not go less than 12/12 with them. Most other species fruit fine with less light; for example, you could try 10/14 or 8/16.

Should I allow my cakes to fruit directly from the perlite?
No, you should use some kind of base underneath your cakes. The lid of the jar your cakes were colonized in makes a convenient base. Otherwise, consider using a piece of aluminum foil or plastic.

My cakes look fuzzy. Is this normal?
Yes, after the cakes are birthed, they will start to appear fuzzy on the surface. If you see any colors besides white, there is cause for concern. Sometimes excessive fuzz, especially around the base of the mushrooms, can mean that the humidity is too high. If the fuzz continues to grow, it could be cobweb mold, which is a contaminant (see page 46).

How long will it take to fruit?
This will depend on many factors, such as the species, the temperature, when they were birthed, and so on. It should not take much longer than 2 weeks to begin to see pins from your cakes. Most will begin pinning much faster, but be patient. The normal time is 1 week for many species. If you don't see any contamination, then you're probably on the right track.

My mushrooms started fruiting in the jar. Is this a problem?
Most likely it is not, but this suggests it's time to go ahead and birth them. If the mushrooms are formed in the jar, consider skipping the dunk and moving them directly into the fruiting chamber.

I've heard of a dunk and roll method of casing a cake. Should I use it?
Some people choose to roll the sides of the cakes in moist, fine vermiculite before putting them into the fruiting chamber. This works just like any other casing. It's something else you can try if you are so inclined.

Some of my cakes are fruiting but others are not. Is there a problem?
Probably not. Just be patient and give the lagging cakes a few extra days.

Harvest

Now that your mushrooms have appeared, it's time to start thinking about the harvest. This is one of the most exciting parts of the process, possibly second only to the time you saw the first pin growing. Each species has a slightly different set of indicators telling you that it has reached its prime. For most mushrooms with a central stem, cap, and gills, that time will be when the gills become visible under the cap. By then, most of the growth has been arrested, and the mushrooms will not get much larger.

When that time comes, simply grasp the mushroom at the base of the stem and gently twist it away from the substrate. The mushroom should easily twist right off. It's not necessary to harvest all the mushrooms on your substrate at the same time; you can harvest them individually as they mature. If the mushrooms have formed in clusters, it's usually futile to try to harvest only part of the cluster. Just go ahead and pluck the whole cluster at the same time.

Aborts

Some mushrooms start off growing as small pins, but never develop into full fruits. This is normal, and these mushrooms are called *aborts*. It's often advisable to harvest, dry, and save the aborts once you harvest your fully mature mushrooms. There are a number of reasons that a mushroom may abort. A sharp change in humidity level or temperature is one of the most common causes. Touching the pins can cause them to abort, as well as spraying them directly with water. I have even had mushrooms abort from excess bleach fumes from cleaning the growing chamber. If many of your mushrooms abort at the same time, you should start looking for a cause, but it's common for a small number of them to abort with each round.

Growers of common edibles will save aborts, but they will usually not include them with the rest of their product, as they are often deformed and ugly. Instead, you can dry and process your aborts into a mushroom powder for soup stocks, if you are trying to retain as much value as possible. Psilocybe cultivators tend to save these mushrooms as well, because they possess higher psilocybin content per weight than fully grown mushrooms.

After all the mushrooms have been harvested from the substrate, it's time to dunk the substrate again. Submerge your cake in cool water for 12 to 24 hours. This will rehydrate your substrate and get it ready to produce the next harvest of mushrooms. After you have dunked your cake, reintroduce it to your fruiting environment.

Drying Mushrooms

Most fungi need to be dried or refrigerated soon after they're harvested. Without taking steps to preserve them, your mushrooms will begin to decompose rather quickly. Cooling them in the refrigerator should allow them to last for a week or so without further action. Always store your mushrooms in paper bags, never in plastic. Mushrooms need to "breathe," and storing them in plastic restricts the airflow, making them decompose faster. If you want to store them for longer than a week, you'll need to dry them.

There are several drying methods that work very well. The most basic is to put them in front of a fan for a day or two. Fan drying will get mushrooms very dry, but if any moisture is left in them, and you try to store them for a longer term, molds may form.

If you're planning to dry and store your mushrooms in a sealed container, you may need to use a desiccant. A desiccant is a substance that removes moisture from the air. It lowers the humidity in the container, removing water from the mushrooms in the process. A desiccant will get your mushrooms cracker dry, allowing them to be stored for longer periods of time. Most people fan dry mushrooms first, and then add them to a desiccating chamber to finalize the drying process. A commonly available desiccant is a product called DampRid, which can be found near the laundry detergents in most large stores.

Properly stored, cracker-dry Shiitake mushrooms will keep for years without losing flavor or quality.

Making a Desiccant Chamber

Materials

- Large plastic bin with a tight-fitting lid and screws inserted at strategic points, to hold the layers of screen
- DampRid or other desiccant
- Paper towels
- Two pieces of metal screening, any size mesh, cut to fit inside the bowl

1 **Start with semi-dry mushrooms.** Fan dry your mushrooms to remove most of the moisture. This will make your desiccant more effective and longer lasting.

2 **Set up the first level.** Using the screen, create two levels in your bowl. The lower level will hold the desiccant; the upper level will hold the mushrooms. Place a piece of paper towel on the lower level and place the desiccant on the paper towel.

3 **Add the second level and lid.** Put the second screen into the container. Place the partially dried mushrooms on the screen. Keep the mushrooms in the chamber until they are cracker dry.

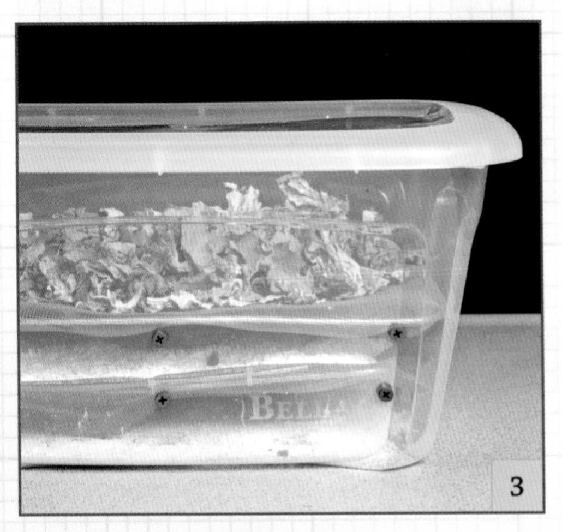

Creating Spore Syringes

Once you've completed your first grow, it's time to start thinking about creating your own inoculum. By propagating the culture yourself, you'll no longer be dependent on inoculum purchased from an outside source.

The easiest option for beginners is to create a *spore print*, which is thousands of spores from a single species of mushroom collected for future propagation. Spore prints have many uses. You can inoculate liquid cultures or agar petri dishes with them. You can also use them to create your own spore syringes. Other options for creating your own inoculum are liquid cultures (see chapter 8) and agar cultures (see chapter 11), but creating spore syringes is a good place to start.

The first step to creating your own spore syringe is to create a spore print. If you cut off the cap of a mushroom after harvest and place it on a piece of aluminum foil or glass for several hours, spores will drop from the gills and collect on that base. Spores obtained this way can be saved for many years. The main goal for you as a cultivator is to create a "clean" print from which you can make your own spore syringes that will be free from any external contaminants.

Spore prints can be saved for many years with essentially no upkeep. This makes it a great method for long-term storage of your mushroom species. The next step is to use the print to create your own fresh inoculum. This is just one of many possible ways to create spore syringes. After you're familiar with the process, feel free to adjust this method to fit your needs. As with inoculating jars, this is a process that you should work through quickly, but not sloppily. Be efficient in your movements and processes. Maintaining sterility in this process is essential, because it will affect every other procedure you perform in the future.

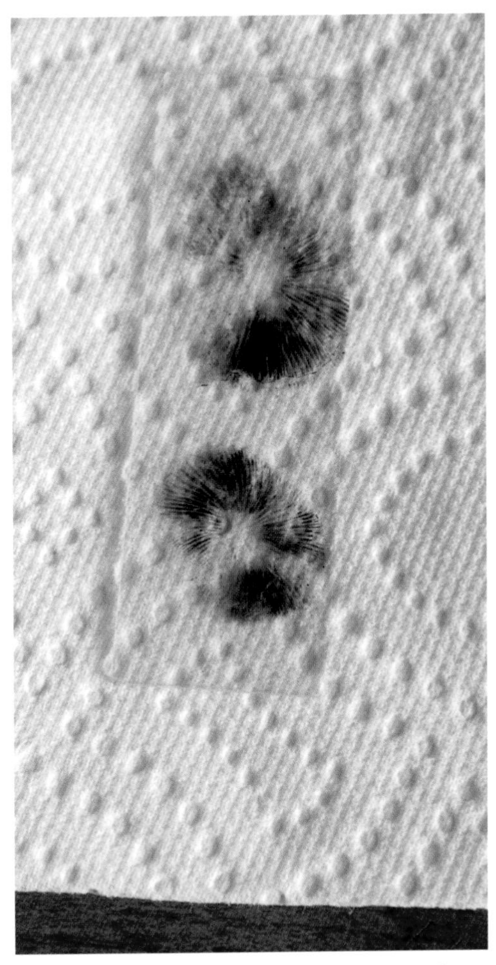

These mushroom spores are ready to transfer from the slide to a syringe. Learning to prepare your own spore syringes eliminates your need to purchase inoculum.

Making a Spore Print

Materials

- Recently harvested mushroom
- Thick aluminum foil or glass slide
- Rubber gloves
- Scalpel
- Small juice glass
- Alcohol swab

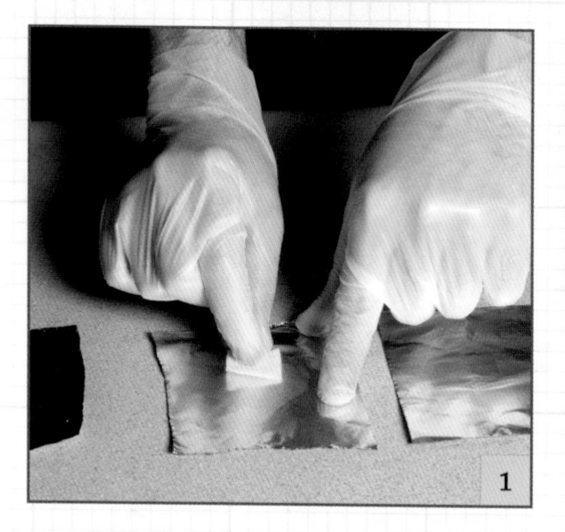

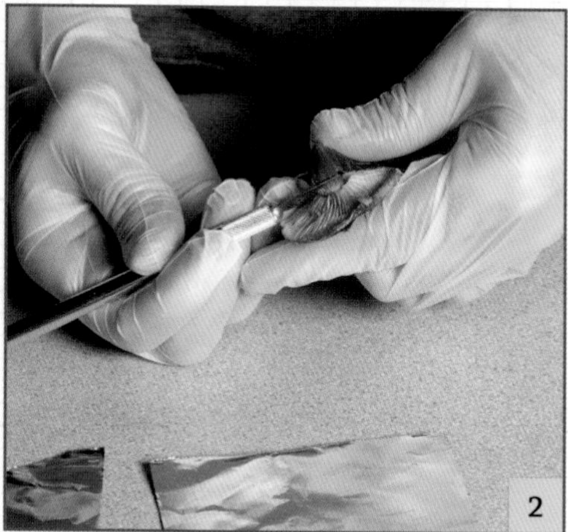

1 **Set up the site.** Find a still-air environment. The inside of a glovebox or a clean cabinet will work nicely. Place the foil or glass into the still-air environment. Wipe the surface down with the alcohol swab. Allow the alcohol to dry for 10 minutes or so with the juice glass over the top.

2 **Harvest the mushroom.** Wearing sterile gloves, harvest a mushroom. Cut off the mushroom cap near the gills, without touching the gills. Place the cap on the foil.

3

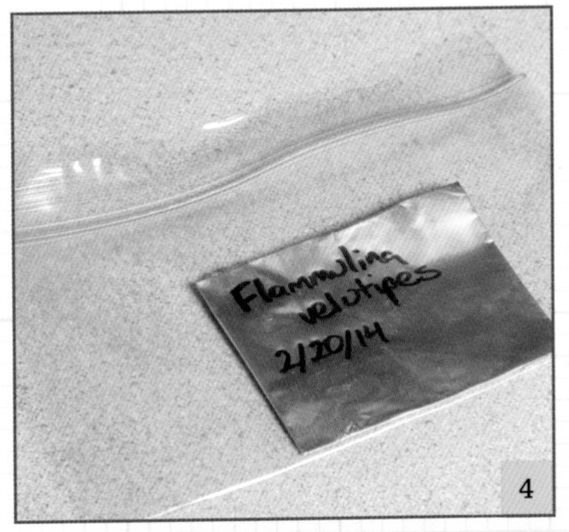

4

3 **Put the juice glass over the mushroom cap.** This will help prevent contaminants from landing on the spore print. (It helps to place a small, folded piece of paper or something similar under one side of the cup to raise one side a fraction of an inch off the surface. This will prevent a seal from forming and allow some humidity to escape.)

4 **Preserve the spores.** Allow the mushroom cap to sit on the foil for 12–24 hours. The spores will drop from the cap onto your base over this time and stick to the foil or glass. Place your print into a plastic bag with a zip lock. If the print was taken on foil, you may choose to fold the print in half and fold over the sides to seal it up. Label the plastic bag using a permanent marker. Record the species and date.

Making a Spore Syringe

Materials

- Spore print
- Glovebox or flow hood
- Empty 10 or 20 cc syringes
- Lighter or alcohol lamp
- Two half-pint (250 mL) canning jars
- Scalpel

1 **Sterilize the tools you'll use.** Wrap your syringe, scalpel, and one empty canning jar in aluminum foil. Fill the other canning jar with water. Pressure cook all items, including the jar of water, for 20–30 minutes at 15 psi. If you don't have a pressure cooker, steam sterilize them in a large pot for 45 minutes.

2 **Fill your syringe.** After the items have cooled, move them into your glovebox or in front of your flow hood. Remove the aluminum foil from around your empty jar and syringe. Open your water jar. Draw water from the jar into your syringe. Put the cap back on your syringe.

3 **Transfer the spores.** Open up your spore print. Using the scalpel, scrape some of the spores off the print into the empty jar. Most of the spores will be floating on the surface of the water. A good amount is about one-eighth of the total print. Adjust the amount according to how dark you want your syringes to be.

4 **Mix the spores and water.** Inject water from the syringe into the empty jar. The spores will become suspended in the fluid. Draw the water back into the syringe. You should now be able to see the spores in the water of the syringe. I suggest expelling the fluid from the syringe into the jar and drawing it back up a second time before finalizing the syringe. This will capture more of the spores in the glass and break up clumps of spores. Finally, put the cap back on the syringe. Repeat this process for all syringes you're making. If your prints are more than a week old, let the syringe sit for 12–24 hours before inoculating jars with it. You can use the syringe immediately if the print is fairly fresh.

PART 2

Intermediate Methods

CHAPTER 5

PRESSURE COOKERS AND FLOW HOODS

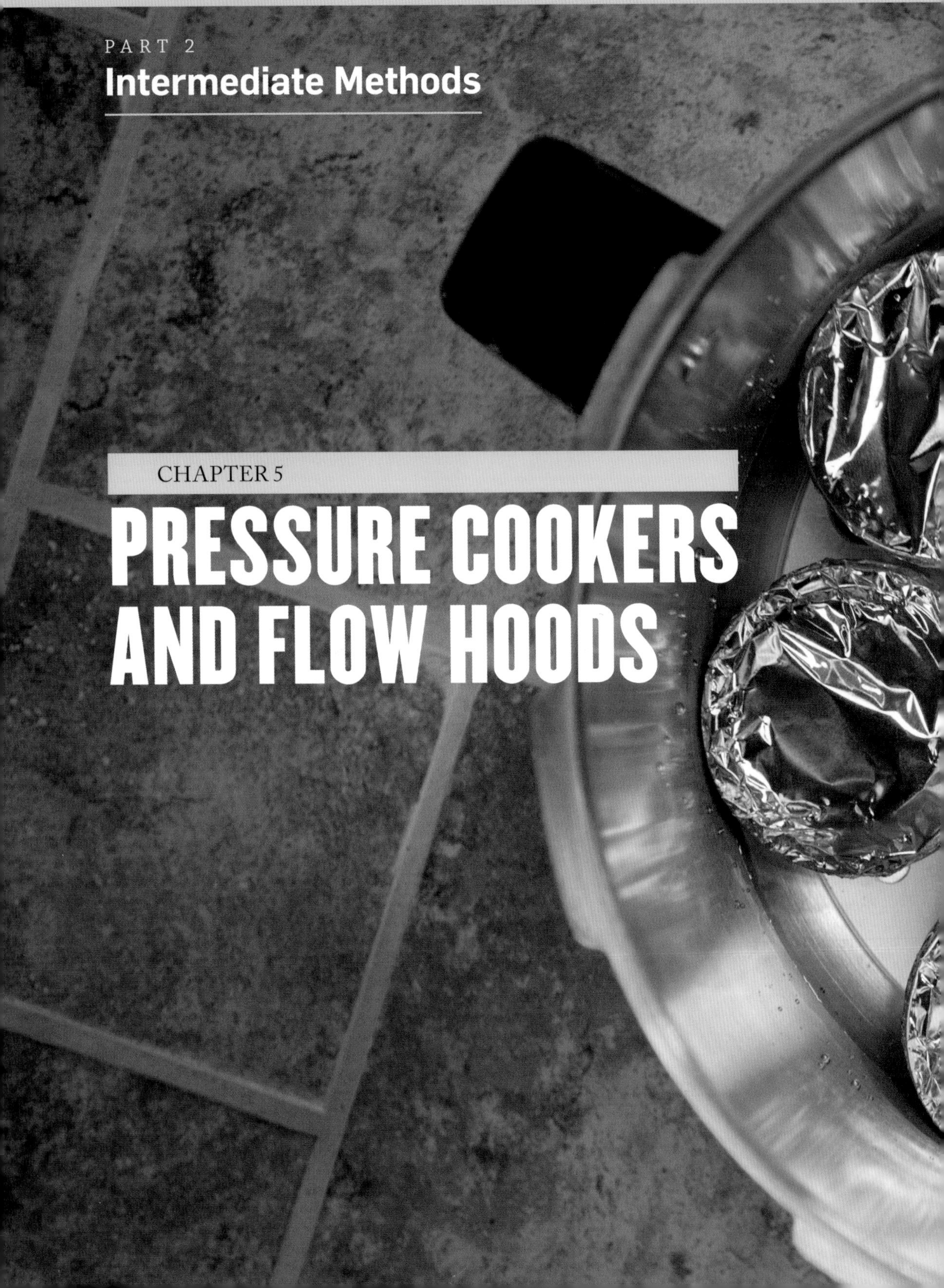

If you're a beginner, PF Tek is a great method because it allows you to grow mushrooms in small quantities without investing too much time, effort, or money. If you're happy with your PF results and don't feel the urge to move on, that's perfectly acceptable. But if you have the itch to expand your mushroom production after completing the entire PF process two or three times, it's time to explore some new equipment and techniques.

One of the main limits to how much you can grow is the amount of substrate you can reliably produce. While it's possible to sterilize and colonize large numbers of half-pint PF jars, it's not a very efficient process, and large increases in production do not come easily. To generate large quantities of spawn, and thus dramatically increase your yields, you'll need to be able to sterilize larger amounts of substrates efficiently.

As you expand your exploration of mushroom cultivation, you should consider two key investments: a pressure cooker and a flow hood. Each of these pieces of equipment will save you time, money, and headaches. Both are well worth the purchase price.

Pressure Cookers

From this point forward, sterilizing substrates will be one of the procedures you perform most often. Pressure cookers play a primary role in substrate sterilization. A pressure cooker is a sealed vessel that heats up the water it contains. As the water heats up and boils, the steam that is generated cannot escape, so pressure begins to build. You may remember the ideal gas law equation from basic chemistry: $PV = nRT$. It means that if the volume (V) and the amount of liquid (n) remain constant, an increase in pressure (P) results in an increase in temperature (T). This means that, at standard atmospheric pressure, there is a maximum temperature boiling water can attain without evaporating into the atmosphere (about 212°F/100°C). Most pressure cookers are rated for operation at up to 15 psi of pressure. By increasing the pressure, these cookers allow you to boil water at higher temperatures, sterilizing your substrates more quickly than would be achievable by other, non-pressurized methods. A pressure cooker operating at 15 psi boils water at around 250°F (121°C).

Prioritizing Knowledge over Equipment

To grow mushrooms well, it's not necessary to spend a lot of money on fancy equipment. Spending thousands of dollars on the best pressure cookers and flowhoods will not guarantee success if you have not taken the time to learn the basics of cultivation. It's important to not only follow the procedures, but also to have a working knowledge of why each step is needed. The best cultivators are the ones who commit to the most learning, not the ones who buy the most expensive equipment.

Pressure Cooker Safety

Pressure cookers can be dangerous if they're not maintained and used properly. I've read reports of cookers exploding during use, so I always treat them with respect, and you should, too. Most of the problems I've read about have occurred with models older than 20 years that lack modern safety features. Avoid second-hand models that may have been dropped, dented, or misused. While new cookers are expensive, you don't want to take the risk of operating a suspect unit in your home.

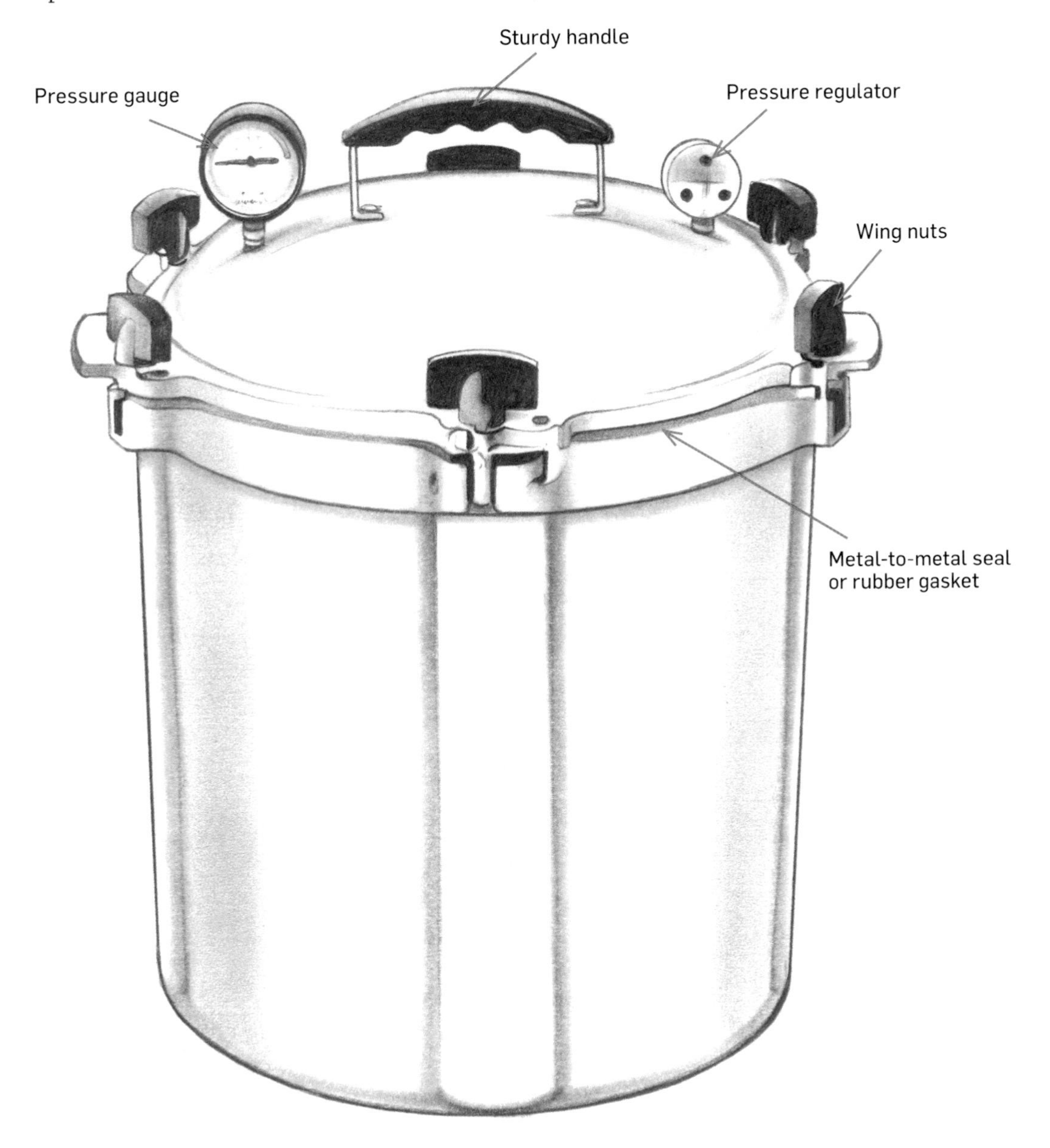

Tips for Safe Operation of Pressure Cookers

Read the instructions. Improper operation is a real threat to your safety, so it's imperative that you know everything possible about your unit. Every unit operates differently. Even the same brand changes from year to year, so stay apprised of all the features of the model you have purchased.

Make sure the cooker has enough liquid. Running the cooker dry can ruin the contents and potentially damage the cooker. To make sure it always has enough water, I stack ring bands underneath the bottom standoff plate. This increases the volume of water in the cooker without bringing the water level too near the tops of the jars. For larger cookers, I often stack two levels of ring bands under the standoff plate if additional water is needed. You can use other items besides ring bands, but the goal is to raise the standoff plate an inch or two while the unit is in operation.

Make sure your jiggler is not stuck to the vent pipe. Most manufacturers recommend that you apply your jiggler weight to the steam vent only after steam has been coming through it fairly steadily for several minutes. If you begin to cook with the weight on, it will sometimes stick to the lid and not become unstuck until the pressure gets to 20 psi or higher. If this happens, the jiggler could blow off the unit entirely once it breaks free.

Release the pressure safely. Let the pressure decrease from the cooker naturally. I used to pour cold water on the hot cooker or soak it in water to speed up the cooling process, but this isn't the safest method. It's recommended for some models, but others expressly recommend against it.

Never move or open the cooker until the pressure is all the way down. If the pressure has not diminished, a large burst of steam or a small explosion may occur. Steam causes serious burns. Be patient.

Always stay near the cooker while it's in operation. Consistently monitoring the cooker is one of the best ways to ensure safe operation. Most of my close calls with the cooker have occurred when I was distracted while the cooker was in operation. The cooker's pressure built unimpeded until it popped the release valve and let steam escape. If this goes on for too long, the cooker will run dry, ruining the contents and potentially ruining the cooker.

Keep kids and animals away from the cooker while it's in operation. This one is self-explanatory.

Maintain the cooker. Follow all manufacturer instructions for maintenance. For Presto and Mirro cookers, this means making sure the gaskets don't need to be replaced. For All American cookers, apply a small amount of petroleum jelly to the metal-to-metal seal. Be sure you know the needs of your individual cooker.

Never put jars or bags directly on the bottom of the cooker. Always use the standoff plate that came with the cooker. The contents will burn without the standoff plate.

Keep your cooker clean. This will ensure proper operation and extend the cooker's life expectancy. Clean both the interior and exterior of your unit on a regular basis. Use a toothpick to clear the steam vent of debris before operation. If you're using a unit with a rubber gasket, remove the gasket and clean underneath it as well.

Make sure all the water is out of the cooker when it's not in use. It's no fun to find several inches of nasty water sitting in the bottom of your cooker, and it's a potential source of contaminants in your lab room. This is especially true if you just cooked agar or liquid culture, and a small amount of nutrients found their way into the water. Give your cooker a quick rinse after every use, and make sure all the water is out of the cooker before you store it.

Do not store the cooker with the lid on. Any moisture left in the cooker while it's idle will safely evaporate if the lid is left unsealed.

A pressure cooker can be useful for sterilizing equipment and small quantities of substrate material.

Heating Pressure Cookers

The largest cooker you can use on an indoor range is 23 quarts (22 L). Using a cooker larger than this on a stove may damage the burner surface because of the weight of the cooker and contents. The cooker would also take an extraordinarily long time to heat up. If you fill up a 41-quart (39 L) cooker with jars, it may never reach 15 psi on the stove. A cooker such as this one, filled with six grain bags, weighs around 90 pounds and will easily damage a stovetop.

You have two heating options to consider when using larger cookers. The first is an electric hot plate; the second is a propane burner. The first question to ask with either of these options is if the heater can support the weight of the cooker.

Small, single, stove-style hotplates, available at big box stores, will not be up to this task for large cookers. They're too small to support the weight of large cookers, and they're rarely rated at more than 1,000 watts. They'll work fine for cookers that hold less than 20 quarts (19 L), but the burners will need to be rated at 1,500 to 2,000 watts to heat the larger cookers in a reasonable amount of time. Look for models that are designed for use in commercial kitchens or hotplates designed for lab use to be sure they'll be able to get the unit up to pressure and support the weight.

The most common time to find propane burners in stores is around Thanksgiving, when people use them to cook turkeys outdoors. These are rated using British thermal units (Btu); the burners I use are rated at around 60,000 Btu. This burner rating generally represents the highest energy output the burner generates. Buying a burner that exceeds 60,000 Btu will not yield much benefit. These burners are usually designed for outdoor operation, so be careful where you place them, and follow the manufacturer's recommendations. Each 15-pound (7 kg) canister of propane, used on a burner of this size will give you 15 hours of actual cooking time.

Stuck Lids

Sometimes the lid of the pressure cooker will become stuck on the unit. Each manufacturer has different suggestions for dealing with this situation. When the lid on my cooker gets stuck, I first remove the jiggler from the unit. A vacuum created within the cooker can hold the lid in place, and removing the jiggler will release this vacuum. Keep in mind that doing so will suck in unsterile air, so you'll want to release it in a clean area.

Often the cooker lid remains stuck even after the jiggler is removed. The manufacturer may suggest that you use a large screwdriver to pry the lid gently near the wing nuts. If you do, be sure you don't dig into the unit with the screwdriver and damage the metal-to-metal seal. I usually keep a hammer near my flow hood (the sterile location where I open the cooker), and will *gently* bang the side of the metal wing nut location until the lid of the cooker releases. One way to prevent All

American cooker lids from getting stuck is to smear a thin layer of petroleum jelly around the inside of the pot where the lid touches the bottom portion of the pot — the site of the metal-to-metal seal. Maintaining this petroleum jelly layer will help prevent sticking.

The wing nut clamps on All American cookers also have a tendency to become difficult to unscrew. A few gentle taps with a hammer can also help to loosen these. Be careful not to hit them too hard, as the plastic side of the wing nut can break off.

Adventures with Pressure Cookers

Every time I've had a scare with a pressure cooker, it's been my own fault. I've never had one explode; thankfully, modern cookers have pressure-release valves that open once the pressure reaches a certain point. Instead, all my negative experiences have been the result of leaving the pressure cooker unattended for too long.

I use an All American 941 for most of my sterilizations. This large, 41-quart (39 L) cooker requires a propane burner or some other high-output heat source to operate. It doesn't take more than 15 or 20 minutes for the cooker to go wildly out of control. The series of events usually goes like this: You set up the cooker and get the flame going on full to heat the cooker quickly. You head out to the lawn to rake some leaves, fully expecting to check on the cooker's progress in five or ten minutes. At this point, a neighbor comes by wanting to talk. Once the conversation ends 15 minutes later and you begin walking back to the house, something registers in your mind and your walk turns into a run. By the time you get back to the cooker, the jiggler has been blown off the top, the pressure in the unit is reading between 25 and 30 psi, and steam is gushing from the valve. Even though you scared yourself, you can still turn down the heat, salvage the interior contents of the cooker, and proceed as if nothing happened.

A similar series of events can happen with smaller cookers on your kitchen stove. Once I got distracted for an hour or more. When I returned to the cooker, there was no steam gushing out. The jiggler had blown off, and the second rubber gasket release had also blown out. All the water had boiled away, and the stench of burnt plastic and birdseed filled the air. The bags I had been cooking had melted to the sides of the cooker, and the interior contents were significantly burned. There was no permanent damage to the cooker apart from melted plastic "scales" on the interior surface, but the next several times I cooked spawn bags, the fresh bags stuck to the interior walls when I was pulling them out after a successful sterilization. There must have been small amounts of plastic left on the walls that melted again and fused to the new bags I was sterilizing. I only lost a couple of bags, but it was an event no one would want to repeat. So don't let yourself get distracted!

FAQ for Pressure Cookers

Can the sides of the jars touch other jars, and can they be stacked while cooking?

It's common practice to stack jars and bags in the cooker. You'll experience no ill effects from bags or jars being in contact with each other.

Should I fill the cooker with cold or hot water?

I usually fill the cooker with cool water. The contents of the cooker then warm up more evenly, allowing the interior contents of jars to maintain high temperatures for the appropriate amount of time.

Can I use water to cool the cooker faster?

There are several things to consider here. As a pressure cooker cools, it sucks in unsterile air. High-speed cooling can bring in this air at a high rate. High-speed cooling can also cause jars to crack or bags to burst from the quick change in pressure. This is less likely to happen with jars, but happens often with bags. High-speed cooling may also cause microfractures or warping in the aluminum body of the cooker. Refer to the manufacturer's recommendations.

Can I cool jars in the fridge after I finish cooking them?

It's possible for warm, sterile jars to suck in unsterile air as they cool. Thus, letting the jars cool in the pressure cooker is the best way to limit contamination. I don't open the cooker until I'm ready to use the contents. I don't recommend removing the jars from your cooker early to cool them in the fridge.

Where should I open the cooker?

That depends on where you have set up your work area. A clean bathroom, a cleanroom, or in front of a flow hood are a few possible locations.

When should I start the timer on the cook time?

Once the cooker has reached 15 psi, you can start the timer.

Will jars ever crack in the cooker?

This happened to me only one time, when the cooker ran out of water. It's more common for jars to crack when you're breaking up the contents after full colonization.

Should I test the cooker before I use it for the first time?

This may be wise, as it lets you become familiar with your unit before you risk compromising the substrates you hope to use. It also allows you a quick test for leaks. When you get a new cooker, fill it with an inch or two (2.54 to 5.1 cm) of water and bring it up to pressure with nothing in it.

Is it okay to release the pressure from the vent?

Yes. Some cookers even have a quick pressure release.

Where can I get a used pressure cooker tested for safety?

Contact your local county extension office.

How do I pressure cook scalpels and syringes?

Wrap them in foil and set them in the cooker to sterilize. Most syringe plungers can be pressure cooked repeatedly without melting.

Why do people put foil over their jars before cooking?

This helps minimize the amount of moisture buildup on the filter. It also helps keep the filter sterile after you pull the jars out of the cooker. This is important if you're inoculating with a syringe.

Can pressure cookers be used on glass stove tops?

Check your manufacturer's recommendations. Usually, the answer will be no. I have had the glass top of a stove crack from the weight of a pressure cooker. This style of stove is not designed to support heavy weights.

My cooker seems to be leaking steam. How do I stop it?

This will depend on your brand of pressure cooker. First, check your instruction manual for recommendations. For Mirro and Presto cookers, it probably has something to do with the rubber gasket. For All American cookers, the lid may not be on evenly. You can also try putting some petroleum jelly on the metal-to-metal seal.

My cooker does not get to pressure very quickly. What's the problem?

The most likely cause of this problem is that the heating element doesn't contain enough energy to heat up the cooker. Use a larger heating element. Also, check for any steam leaks from your cooker.

The filters of my jars get wet while cooking. Is this normal?

Yes, this is normal and is nothing to worry about. The entire interior of your cooker will fill with steam during the sterilization process, so some moisture is unavoidable; however, you do not want any standing water in the jars or bags.

Is there any way to sterilize jars without a pressure cooker?

Yes. There is a process called Tyndalization, or fractional sterilization. I haven't mentioned it because I generally don't like this process. It involves boiling the jars on three consecutive days. The results are often unreliable, many people have problems with the resulting grain texture, and it is highly inefficient. You can research this process online if you need more information.

My spawn bags burn through. Why does this happen?

There are several causes for this. The first is heating up the pressure cooker too quickly. The bags change consistency as they heat up, and if this happens too quickly, they can develop holes. The solution is to minimize sharp changes in temperature or pressure while cooking. This also applies to sharp changes in pressure and temperature during the cool-down process. You can also try wrapping a towel or something similar around the bags to prevent contact with the metal sides. This should only be necessary as a last resort.

Flow Hoods

Gloveboxes work fairly well for many basic sterile techniques in mushroom cultivation, such as making syringes, simple agar work, grain-to-grain transfers, and other simple procedures. I used a glovebox for my first 3 to 6 months of mushroom cultivation. But if you plan on sticking with the hobby for a while and want to try some of the more advanced methods, you'll need a laminar flow hood.

A flow hood is essentially a large box that sits on a table with a high-efficiency particulate air (HEPA) filter facing you, and a fan forcing air down through the filter. This setup creates an area of sterile air directly in front of the filter, and allows you to perform sterile procedures with unrestricted movement in an open-air situation. The goal of these filters is to create a laminar flow, a series of air currents emitted straight out of the filter across the work surface, without mixing with other side currents.

Although they can be quite expensive ($300 to $600 to make one yourself; $600 to $1,600 or more to buy), a flow hood pays for itself many times over in terms of time saved and contamination averted. I cannot emphasize enough how much easier mushroom cultivation is if you have a flow hood at this point in your education. You can probably find your way through most of the processes in this intermediate section without a hood, but I'm sure you'll start to see the benefits of one as you begin to work through these processes without it. If you have some construction skills, it's fairly easy to build one yourself.

Let's talk about some of the considerations you should be thinking about when making or using flow hoods.

A flow hood with a HEPA filter creates a flow of sterile air, which will allow you to perform sterile procedures directly in front of it.

Size

Size is the first consideration when looking into HEPA filters. My first filter was 12 inches (30 cm) high and 24 inches (61 cm) wide. It worked very well for a year or two, but when I was ready to increase production, the unit wasn't tall enough to transfer grain jars efficiently into large spawn bags; however, I recommend starting with the 12" × 24" (30 × 61 cm) hood, especially if you're still testing the waters with the hobby. You can easily perform grain-to-grain transfers in quart jars with a hood of this size, and it's still fairly portable. It's also less expensive, and if you decide to continue growing mushrooms, it makes a great second hood for smaller projects.

If you intend to produce a significant amount of spawn in bags, or if you want to make sawdust block cultures of species like Shiitake or Lion's Mane, you'll need a larger hood. The one I've been using for years is 36" × 18" (91 × 46 cm) and has been sufficient for producing significant quantities of spawn over the years. You won't need a larger one unless you're planning a significant commercial operation.

Efficiency

If the filter you're considering is not a true HEPA, which offers 99.99 percent efficiency for particles down to 0.3 microns, continue your search elsewhere. There are several mushroom companies that sell filters from the HEPA Corporation, and whose filters are designed for mushroom cultivation. This is where I've always acquired my filters. These filters usually have aluminum frame construction, and are about 8 inches (20 cm) deep.

CFM Rating and Blower

If you're considering building your own flow hood, your first task, after acquiring the proper filter, will be to match a blower to the filter size. These blowers are known by several names, but "squirrel cage blowers" or "HVAC blowers" are some common ones. They're commonly used for many different HVAC-related purposes, but also work well for flow hoods. Most of the smaller blowers come with a mounting bracket attached to the blower, but larger cubic feet per minute (CFM) blowers may require you to purchase a mounting bracket separately.

If you look online or in books, you'll find many different calculations related to matching the blower to the hood size. I'll try to make it simple: consider a blower of between 200 and 250 cfm per cubic foot of filter. This assumes a 0.8 static pressure HEPA and a 0.2 SP pre-filter.

Filter Size	Blower Size
12" × 24" (30 × 61 cm)	400–500 cfm
24" × 24" (61 × 61 cm)	800–1000 cfm
36" × 24" (91 × 61 cm)	1200–1400 cfm
48" × 24" (122 × 61 cm)	1600–2000 cfm

Building a Flow Hood

Building your own flow hood is a fairly simple process, even for those with rudimentary carpentry skills. You're really just making a wooden box of the proper size and attaching a blower to a hole cut in the top. Here's the process for a 12" × 24" (30 × 61 cm) hood.

Materials

- 4' × 4' (1.2 × 1.2 m) sheet of ¾" (2 cm) oak plywood
- 1" × 1" (2.5 × 2.5 cm) furring strip, 8' (2.4 m) long
- 1" (2.5 cm) edge trim, 8' (2.4 m) long
- 1½" (3.8 cm) wood screws
- 1" (2.5 cm) nails
- Roll of ⅜" (9 mm) weather stripping
- All-purpose silicone sealant

Equipment

- 400–500 cfm blower
- 12" × 24" × 5⅞" (30 × 61 × 15 cm) HEPA filter, efficient to 0.3 microns
- Power strip with on/off switch

The finished flow hood box is ready for the filter to be installed.

Begin by cutting out the shape of the box. I will present some sample measurements, but if the total size of your filter is not exactly 12" × 24" (30 × 61 cm), you will need to make some corrections.

Top and Bottom – 25½" × 16¾"

2 Sides – 16¾" × 13½"

Back – 24" × 15¼"

Assemble your box. Begin by drilling pilot holes, which make it easier to insert screws and less likely that you'll split the wood.

Measure the exact depth of your filter. Install the furring strips that distance, plus the thickness of the weather stripping, back into your box. This will be the barrier that holds the filter in place from the back.

Install the weather stripping facing out the front of the box. This will ensure a tight seal.

Measure the opening of your blower, and cut a hole of that size into the top of your box between the furring strips and the back.

Use silicone sealant to seal all the joints on the backside of the box. This will ensure that it's airtight. Lay a small bead of silicone around the blower opening.

Install the blower and screw it in.

Slide your filter into the front opening. I tack several nails into the sides to hold it firmly in place.

Cut your front trim pieces to give the box a finished look.

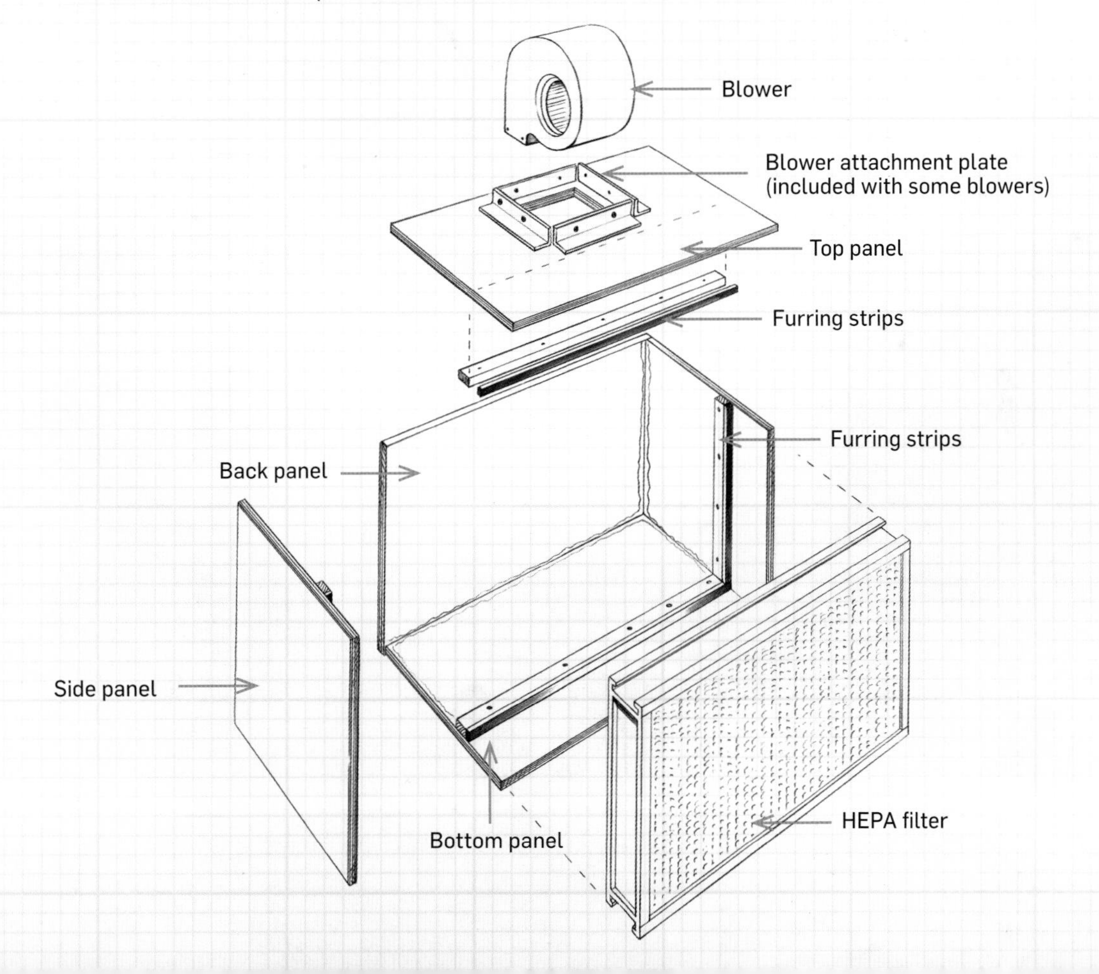

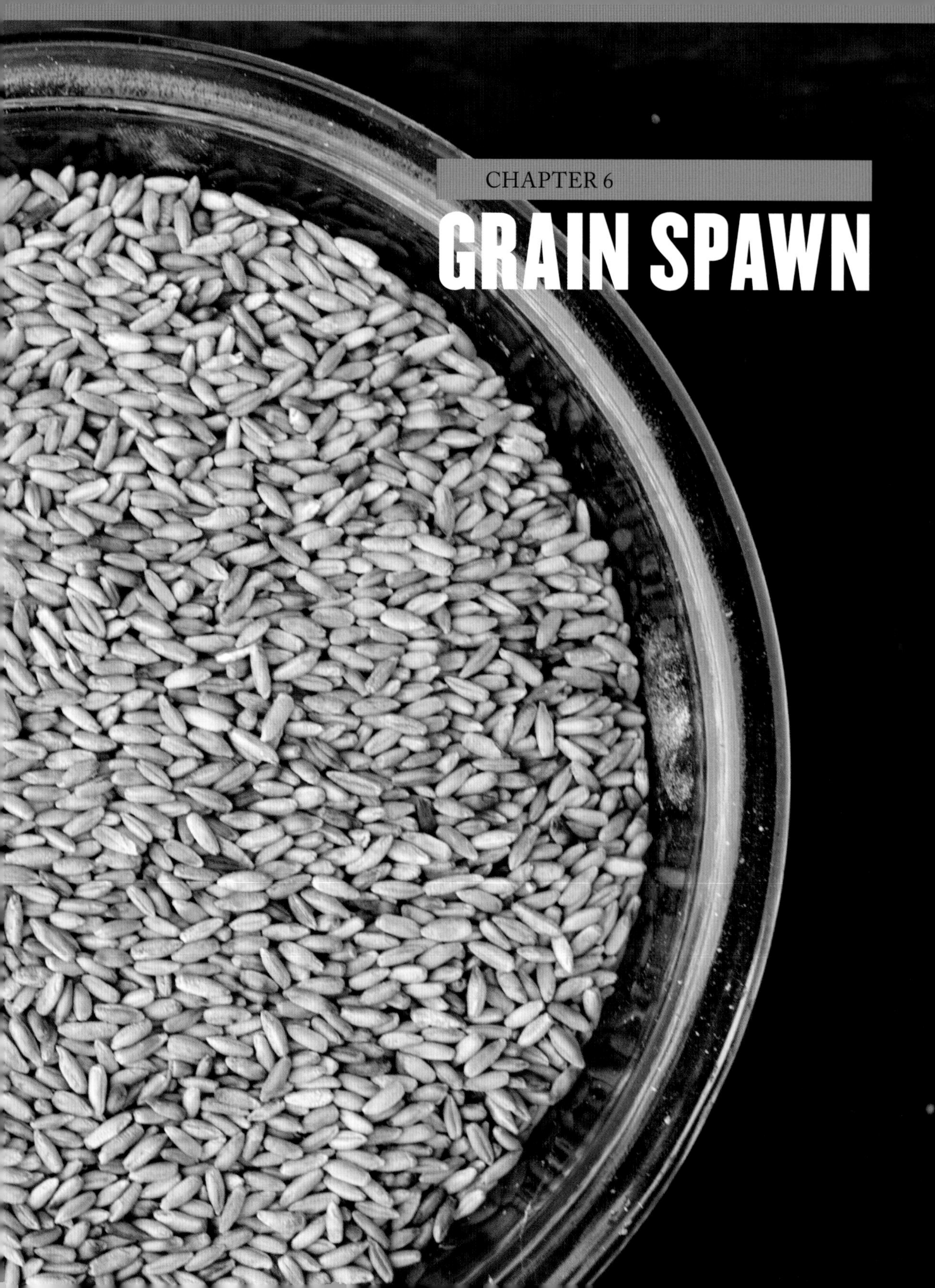

CHAPTER 6

GRAIN SPAWN

The next step in your mycological education is the generation of grain spawn. Grain spawn is used to propagate a sterile culture of mycelium that can either be fruited itself or transferred to other substrates as spawn. Generating grain spawn allows you to expand the size of your mycelium culture and increase your final yields. Many types of grain can be used successfully for spawn.

An Overview of the Process

Grain spawn generation involves a number of basic steps. First, you'll choose the jars to use and prepare jar lid filters, which allow air into your jars during colonization. Next, you'll have to choose a grain. Any grain you choose requires similar processing, but each has individual characteristics that you should know about before you decide on one over another.

Because most grains have a hard outer shell and low water content, you'll have to hydrate the grain before you use it. After hydration, the grain must be sterilized in a pressure cooker. Up to this point, you could get by without using a pressure cooker, but the procedures in this chapter require one.

After you fill your jars with hydrated grain, inoculate them and allow them to colonize. Finally, you'll have to decide how to use your grain spawn. You can, for example, make a grain-to-grain transfer, transfer the spawn to a bulk substrate, or make a casing. Grain-to-grain transfers are covered in this chapter (see page 115); bulk substrates are discussed in chapter 13, and casings are explored in chapter 7.

Choosing Jars

Quart canning jars are the primary containers you'll use when working with grains. There are two main varieties: regular mouth and wide mouth. The only difference between these two options is the size of the opening. I tend to use both sizes with some regularity, and pick up a couple dozen of one kind or another whenever I see them on sale. Wide-mouth quarts tend to be easier to fill when doing your initial prep work, as the entire head of a standard kitchen scoop will fit into the jar. This is not the case for regular-mouth jars. I use regular-mouth jars when doing grain-to-grain transfers because I can pour the grains from one jar to another in a more controlled fashion than is possible with the wide-mouth jars.

Making Jar Lid Filters

Just as you turn the lids of PF jars upside down to allow some air exchange, quart jar lids must also be modified with filters to facilitate air exchange.

Making Holes

The first step in preparing your quart jars for grains is to create a hole in each of your jar lids. While PF jars have four holes, quart jars require only one hole, between ¼ inch (6 mm) and ¾ inch (2 cm) in diameter, placed directly in the center of the lid. Most of the holes I make are around ⅜ inch (1 cm) in diameter. One simple way to make the hole is with a drill bit of the desired size. Most drill bits pop right through the aluminum lids without much hassle.

Another option is to use a large Phillips screwdriver and a hammer to create the hole. With the lid and ring band on the jar, place the screwdriver on the lid and give it a solid whack with the hammer. One good hit should pop it through. This method produces some sharp edges on the underside of the lid, so it's best to take the lid off, flip it over and hammer down the sharp points on a hard surface. If you don't do this, your fingers will regret it: those sharp edges work like razor blades.

Poly-fil Filters

Although not the cheapest or most efficient filter design, Poly-fil filters work remarkably well, and I've used them successfully for many years. Poly-fil is a synthetic cottony material, available at big box stores and craft stores, and is most often used to stuff pillows. One of their main benefits is that you can inject a spore or liquid culture syringe directly though the hole and filter and into your sterilized grain. To inject solution into

Some jar lids are designed specifically for mushroom production and have built-in filters.

jars using any of the other filter methods, you'll need to remove the jar lids completely. If you are transferring from an agar dish, you'll need to remove the lid of every jar for inoculation anyway, but if you frequently work with spore syringes or liquid culture syringes, I recommend the Poly-fil filter, as it saves you time during inoculation and works very well.

There are several varieties of Poly-fil available; the type you need doesn't absorb water. At least one common pillow stuffing material absorbs water and will come out soaked after your grain jars are pressure cooked. A soaked filter nullifies any potential for air circulation in the jar, thus defeating the purpose of creating the filter. It also may allow contaminants to grow through the filter. The Poly-fil brand is made of the proper synthetic polyester and is *hydrophobic*, meaning it repels water. This is the material I recommend.

To make your own filter, first punch a hole in the lid, then stuff it with Poly-fil.

Using Poly-fil to create a filter in the lid of your quart jars is not a very complex process, but you'll need some experience to perfect the method. The trick is to use as little material as possible, but still lodge the filter firmly in the opening.

One other consideration is to make sure that the filter is not packed in too tightly. If it is, air exchange will be inhibited, and you won't be able to inject the syringe directly through the filter. It's a good idea to test your filter using an empty syringe before you pressure cook it, rather than running the risk of wasting your sterilized grain. It may take some time to become proficient at making filters, but rest assured, you'll learn over time.

These filters are reusable, but don't use the same filter more than two or three times, as contamination rates go up the more the filter is reused. Always be sure that the filter still has a snug fit before reusing it. If you have any question about whether to reuse it, just go ahead and make a new filter. They are inexpensive and don't take very long to make.

Making a Poly-fil Filter

1 **Grab a handful.** Tear off a piece of Poly-fil that is larger than a golf ball, but smaller than a baseball.

2 **Compress and stuff.** Roll the ball in your hands to create a compressed sphere of stuffing. After it's compressed, twist it a few times, place it over the hole in the lid, and pull it through the underside. There should be enough compressed Poly-fil to fit snugly in the hole. Remember that you'll have to shake the jar somewhat violently in the future, and if the filter pulls out easily, your jar will become contaminated and you'll have to throw it away.

3 **Cut off the excess.** When your filter fits snugly, give it a gentle pull. If it doesn't move much, you have successfully made a jar filter. If the filter hangs down more than 1" (2.5 cm) through the hole, cut off the excess with scissors.

1

2

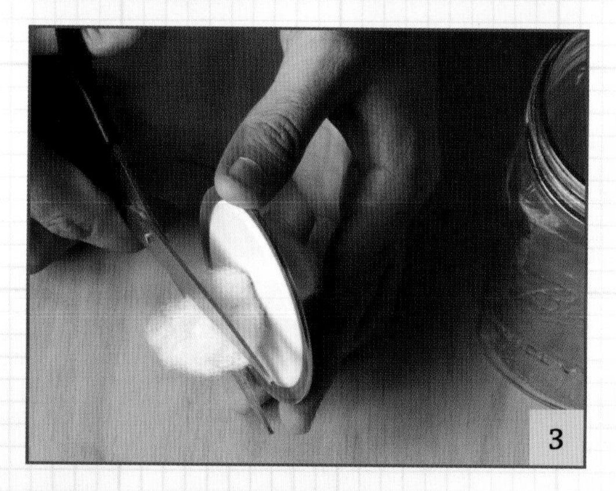

3

Tyvek Filters

Tyvek is another filter option that many growers use. This is a breathable paperlike plastic material used in many lab-related products, such as cleanroom suits. Instead of stuffing it through the hole in the lid, you can layer this material over the surface of the jar, under the lid or between two lids, and secure it with the jar's ring band. Tyvek's major drawback is that you cannot inject a syringe of liquid culture or spores directly through it; you must completely remove the lid for injection. This increases the potential risk of contamination, but Tyvek is a good choice for a filter if you are using the jars for grain transfers.

If you want to see what Tyvek looks and feels like, go to your local post office and look at Priority and Express Mail document mailers, some of which are made of a thin form of Tyvek. This weight will need to be double layered on the jar, because contaminants can find their way through a single layer. You can find sources for Tyvek online. Some types, such as those used to make cleanroom suits, are thinner and more flexible than the mailers. Some painter's suits and house wraps are also made of Tyvek.

To make a Tyvek filter, cut two circles of the material and place them between two lids that have holes cut in the same spot. This creates a double-layered permanent filter between the two lids, because the lids will seal together after they are pressure cooked the first time. Extra lids are available wherever canning jars are sold.

Synthetic Filter Discs

Another option, similar to Tyvek, is to use synthetic filter discs. These filters are sold specifically for mushroom cultivation. They are available as precut circles in two sizes — regular mouth and wide mouth — and both are 99.97 percent effective against microbes as small as 0.3 microns in diameter. That makes these filters nearly as efficient as a HEPA filter. They are also reusable and able to withstand pressure cooking numerous times without degrading.

These filters have several drawbacks, the first being cost. They often cost $1 or more each, which is not too exorbitant for the service they provide. I never adopted them on a wide scale because you cannot inject through the filter into the jar. The time you save by not having to create Poly-fil filters is offset by having to open each jar for injection. I would probably use synthetic filter discs exclusively if my primary inoculum were agar sections instead of a liquid culture, as I would have to open the jars with each inoculation anyway. I would also probably use synthetic discs more often if I were doing hundreds of quart jars through grain-to-grain transfers. Ultimately, the filter type you choose should fit the methods that work best for you and the scale of cultivation work you do. That being said, all the filter options discussed here are dependable.

FAQ for Synthetic Filter Discs

You outlined several different jar lid methods. Which would you recommend?
That depends on the ultimate use. If you're injecting jars with a syringe, I would use Poly-fil. If you're doing grain-to-grain transfers, use Tyvek or a filter disk.

Synthetic filter discs are another option; they work well but are expensive.

My jars look very dry after a couple of weeks. Could the filter be the cause?
Yes, that's possible. If you have drilled too large a hole in the jar lid, too much moisture could be escaping from the jar, or too much air could be coming in. If you use just Tyvek, without a metal lid, it will allow too much moisture to escape, drying the jar out. Dried-out jars will not colonize completely.

How often should I replace my filters?
Anytime you have contamination in a jar, throw away the filter. If you are using Poly-fil or Tyvek, replace the filter every three or four cookings. Synthetic filter disks should be reusable as long as there has not been contamination.

My Poly-fil filter is soaked with water after cooking. Is that okay?
It might be damp from the sterilization process, but it shouldn't be wet. If it is, be sure you chose the right type of stuffing.

Can I inject through a Tyvek filter?
Syringe needles have sharp points, so you can often push them through a Tyvek filter. The problem is that you are left with a hole in the filter. This problem can be solved by placing a small piece of micropore tape, used on injuries, over the hole. I can't recommend this method, as it introduces several variables that I don't like, but many people have used it with success.

My Tyvek melted. Why?

Tyvek should be stable up to 275°F (135°C), well within the temperature bounds of a pressure cooker's 15 psi. One of two things could have happened: (1.) The pressure got way too high; or (2.) The jars' ring bands were touching the metal side of the pressure cooker and conducted too much heat directly to the filter. If neither of these is the problem, I would switch to a different type of Tyvek or another filter type.

Does it matter which side of the Tyvek is facing out?

No.

I have seen plastic lids available for canning jars. Will these filters work with plastic lids?

Yes, all of these filter methods should work with plastic canning jar lids. Just be careful, as it's easy to crack them when drilling a hole in them.

Preparing and Inoculating Grain

Now that you've made your filters, you're ready to select and prepare the grain you'll use for spawn. There are four different grains that are commonly used in home mushroom cultivation: rye, millet, birdseed, and popcorn. Each has its own advantages and drawbacks, so your choice will depend upon your individual circumstances. There's no "best" grain for mushroom cultivation.

One of the first considerations is availability. You don't want to drive 20 miles to a specialty feed store when there are alternatives available nearby that work just as well. You probably won't want to order grains online, as shipping costs nearly as much as the grains themselves, effectively doubling the price. And since at least one of the four major grains described here is available in every community, you shouldn't have to order online. Most of these grains are interchangeable in most applications.

Another thing to consider is ease of preparation. Each of these grains uses a slightly different method for hydration and prep. Some are best hydrated in boiling water; others benefit from a 24-hour soak. Your needs may be different if you're processing 100 jars rather than 10, or if you're working in a small kitchen rather than a large garage. Some methods are messier than others; some create unpleasant smells. All of these considerations will be discussed with each grain.

A final consideration for which grain to use is the surface area of each grain. Grain surface area affects the speed of colonization. Small grains, such as millet, fit closely together in the jar; consequently, they have a very large total surface area to colonize. Larger grains, like popcorn, have a smaller total surface area, with many less units in the same volume. If all other factors are equal, the larger surface area will need more time to achieve 100 percent colonization. The differences in colonization times are not staggering, but they can be important in some circumstances. For example, a jar of popcorn will probably reach 100 percent colonization several days faster than a jar of millet.

As with most things in life, there's a tradeoff for faster initial colonization time. Usually, colonized grains in quart jars are broken up and spawned (transferred) to another substrate. Small-kernel grains, such as millet, will break up into many more individual pieces than larger grains, like popcorn, if you start with the same volume of each. These smaller grains offer some advantages for certain methods: more individual colonized kernels, for example, mean more individual inoculation points in the substrate you're transferring to, and ultimately faster colonization of this transfer substrate. The mycelia will not need to travel as far to meet the nearest startup colony. For many of the methods discussed later in this book, such as using bulk substrates, more but smaller transfer colonies will yield faster colonization.

Rye or Rye Berries

Rye is the grain most commonly used for at-home mushroom cultivation. Also referred to as "rye berries," rye is generally ground for flour or used as a field crop. It is the first grain I used for mushroom cultivation and works reliably for growers

These sterilized rye grain jars are ready for inoculation.

of all of the common edible and medicinal species. It has a nice balance of ease of use and grain size, and can be used as a primary substrate for fruiting or as a spawn for other substrates. Its main drawbacks are local availability and difficulty attaining the proper moisture level.

The best places to find rye, if you live in a city, are health food stores and organic groceries, where it's sold by the pound with other bulk grains. If it's not in stock, these stores are usually willing to order it for you and have it available within a week. Typical bag sizes are 25 and 50 pounds (11 and 23 kg). Be aware that organic rye may be the only offering; this usually costs quite a bit more per pound.

If you live in a rural or suburban area, be sure to check out local feed stores. They may not have rye in stock, but they can order it for you; it often comes in 56-pound bushel bags. Rye used as field seed, commonly called *winter rye,* is usually much cheaper than rye from an organic grocery. Be aware, however, that some rye available at feed stores has been treated with fungicides. Treated seed usually costs more, and should be clearly labeled. Many feed stores now offer organic seed as well.

Thwarting Endospores

Soaking your grain for 12 to 24 hours ensures adequate hydration, but it also has another benefit: it allows bacterial endospores to germinate, and thus become more susceptible to sterilization temperatures. Pressure cooking grain is necessary because bacterial endospores can survive temperatures of 212°F (100°C). Sterilization at 250°F (121°C) for 20 minutes is required to kill these endospores. Remember that the entire contents of the jar must reach that temperature for the recommended time. The interior of the jars or bags takes longer to heat up than the outer areas, so actual cooking times must be longer than 20 minutes to ensure proper sterilization of the entire substrate. If you're having problems with bacteria in your grain jars, you may want to try longer sterilization times and/or switching to the soaking hydration method.

Preventing Common Problems

When pressure cooking or boiling any of these grains, be sure to avoid overheating, which can cause grain kernels to burst or explode. When kernels explode, they release starches that can cause your grains to clump and can also make bacterial contamination more likely.

Sometimes rye grain can become stuck together and clumpy after it has been pressure cooked. To prevent this, remove the jars from the cooker during the cool-down process, while the cooker is still warm but the jars are cool enough to handle, and shake them to mix up all the kernels in the jar. Shaking the kernels at this time will keep them from sticking together. Remember that jars suck in air as they cool, so let them finish cooling in front of a flow hood.

Hydrating Rye

Some mushroom cultivators like to combine several hydration methods when preparing rye. You might, for example, follow a 24-hour soak with a 15-minute boil. If you are boiling the rye for this short a time, you can use a higher boil without worrying about kernels bursting. Also, if you are ending your hydration with a heavy boil, it's possible to ignore the final step of rinsing the rye in the sink. Instead, just let it steam dry for 10 minutes or so in the strainer before loading your jars.

Any combination of these methods will work fine as long as they bring the rye to a 50-percent moisture content, there are not many burst kernels, and the individual kernels of rye don't stick together when the process is complete.

Popcorn

Along with rye, popcorn is the grain I use most often in quart jars. Corn's large kernels allow for very fast colonization times, with the jars attaining 100-percent colonization several days faster than a jar of rye. I also find corn very easy to work with in a home-based environment. Because the kernels are large, it's not very messy, and it's easy to clean up if there are any spills. Unlike the three methods included for rye, I've included only one method here for preparing popcorn. There are several others out there, such as a boil, but I find pressure cooking the most reliable, and it allows the jars to be made the same day with no 24-hour hydration period necessary.

One of popcorn's big advantages is its availability. Nearly every grocery and big box store carries large quantities, usually in 2- to 4-pound bags. If you need 25 pounds (11 kg) of corn in a day, it's possible in most communities. The same cannot be said for rye.

Prices for popcorn can vary significantly; organic popcorn, for instance, can cost nearly twice as much as nonorganic. One advantage of choosing organic corn is that, as with rye, you can order a 25- or 50-pound (11 or 23 kg) bag of organic corn at your local health food store. You can also check online for other local sources, since many businesses sell large bags of popcorn. The main city in our state has a "concessions supplier" where you can pick up a dozen 50-pound (23 kg) bags the same day if you need to. There are also popcorn farms, where large quantities can be attained at very reasonable prices.

The main drawback to using popcorn as spawn is the smell during hydration. The method outlined here uses a pressure cooker to hydrate the corn, and wherever you cook will smell strongly like cooking corn. I have grown accustomed to the smell, but some people don't like it at all. Keep that in mind as you consider grain choices.

Wild Birdseed and Millet

For mushroom cultivation purposes, wild birdseed and millet are essentially synonymous. The best birdseed for grain spawn has millet as the primary ingredient, and the processes for both grains are the same. Both grains have very small seed sizes, and are your best option if you plan to transfer colonized quart jars directly to a bulk substrate.

The main benefit of using birdseed and millet is that they produce many inoculation points for bulk substrates. If you wanted to spawn your grain quarts directly to straw or compost, one of these grains would probably be your choice. And birdseed has other benefits, too: it's available in large quantities in every community, it's inexpensive, and it has very little smell when it's cooking.

Be aware that birdseed can be very hard to shake in the jar. It tends to solidify into a solid mass of seed if you don't shake the jar while it's cooling down from the sterilization cycle. Keep this in mind when using this method.

I use birdseed and millet primarily in spawn bags (see chapter 12) because it's one of the most inexpensive grain options per pound, and spawn bags require large amounts of grain.

Grain Hydration

There are multiple ways to hydrate grains. I've highlighted the most popular methods for each grain, but boiling, soaking, or pressure cooking can be used for any of them. The ultimate goal is to get your grains to be around 50 percent moisture by weight; anywhere in the range of 45 to 60 percent is ideal. All the methods described in this chapter should hydrate your grain to that extent without requiring any additional steps.

To calculate the moisture content of your grain, hydrate your grain using any of the methods described. After hydration, weigh a generous handful of hydrated grain using some scales. Write down this number; this is your grain's hydrated weight. Now place that same handful of grain on a cookie sheet and bake it in the oven at the lowest setting until it dries out. Be sure not to burn it, just get it dry. This may take several hours. When dry, weigh it again; this is your grain's dry weight. Now perform the following calculation:

$$\text{Moisture Content} = \frac{\text{Wet Weight}}{(\text{wet weight} + \text{dry weight})}$$

This calculation can reassure you that you're hydrating the grain appropriately. After you work with a particular type grain for a while, you'll be able to tell if the moisture content is right by the size and appearance of the grain.

Rinsing Grain

Most of the grains you will be working with need to be rinsed after hydrating, to remove external starches. These starches tend to cause the grain to be sticky and clump together, making colonization more difficult. The kitchen sink, using the spray hose, is the most common place to rinse your grains.

A second option for rinsing larger amounts of grain is to place strainers over large bins outdoors, in the garage, or even in the kitchen, and use the hose from an outdoor faucet as your water source. I find this to be the easiest way to work with large amounts of grain at home. I set up two or three storage bins with as many strainers as they can hold. The best types of strainers for this method are metal ones with side supports that telescope out, allowing them to sit on the tops of the bins.

Remember that rinsing grain just once in the strainer will not remove all the starch, as the water can't easily reach all sides of all the kernels equally. You'll need to shake the strainer and then rinse it again. This is especially important if you're using popcorn. One easy way to shake up the grain is to put another similar strainer upside down over the one containing the grain and flip it into the empty strainer, then rinse it again.

Another way to shake grain that takes some skill and practice is to flip it into the air as you might flip a pancake. While holding the strainer, move it down, then up, and back toward you with a slight flipping motion. If done properly, the grain on the backside of the strainer will slightly flip out of the strainer and fall on top of the rest of the grain. This technique works quickly and efficiently, but takes a lot of practice if you don't want grain all over the kitchen or garage floor.

Shaking Jars

There are a couple of reasons to shake jars during the grain spawn process: the first is to prevent the grains from clumping in the jars; the second is to speed up the colonization process. Grain is most likely to clump in a jar and needs to be shaken just after it's pressure cooked for the first time. This is especially true if you're using

One way to break up mushroom mycelium — to prevent grains from clumping and to speed up colonization — is to knock the jar gently against a spare tire.

smaller grains, such as rye and birdseed; it's less likely if you're using popcorn. The smaller grains may become hard to break apart if you don't shake them while they're still warm from the pressure cooker. If they're not broken up then, the inoculum will have a harder time evenly colonizing the jar.

There are many ways to break up the contents of a jar. Sometimes simply shaking the jar forcefully up and down is sufficient to break up the interior contents. (Be careful not to displace the filter in your lid during this process!) Shaking by hand isn't sufficient for many grains, however, especially if they're fully colonized. Instead, you'll need to find an object that you can use to make contact with the jar. I find that a tennis shoe or a spare tire works best.

Jar Colonization

After your jars are inoculated, it usually takes a week or two for them to become fully colonized. The amount of time it takes depends on the process you used to inoculate and the mushroom species you chose. During this colonization time, you need to find a warm, dry, dark place for your jars to sit. The ideal temperature depends on the species, but most species do best at 70 to 75°F (21 to 24°C). Colonizing jars can sit at room temperature on a lab shelf and do just fine. You could build an incubator to warm them up to the ideal temperature (see chapter 4), or find a cool place if the ideal temperature is lower than room temperature, but the truth is that, for most species, "ideal" temperatures will speed up colonization by a day or two at most. This might be important if you're on a strict schedule, but for most home growers, it isn't an issue.

While you wait for the jars to colonize, the key thing to keep an eye out for is the appearance of colors other than the mycelium you have injected (see chapter 3). Most mycelium is white, although some common edible species can be black, gray, or even orange. Green is universally bad, because it's usually a mold called *Trichoderma* or a species of *Penicillium*.

If your grains appear to be wet in the jar, it may mean that a bacteria called *Bacillus* has colonized your grain. This is probably the most common problem I've encountered, more often even than mold contamination.

If bacteria have colonized your jar, their growth will inhibit the growth of the mycelium you're trying to propagate. After it has been inoculated, there's little you can do to save a jar that has been infected with bacteria; you'll probably just have to throw it away. One simple way to test for the presence of bacteria in the jar is to smell the filter. If it has been colonized with bacteria, you may detect a sweet or foul smell.

One final tip: shake up your jars a day before you plan to transfer them. If the mycelium in a jar doesn't recover sufficiently during that day or the jar looks odd after shaking, you may have a bacterial contaminant, and you may not want to transfer that jar. Be sure to read chapter 3 on contamination, where these issues are discussed thoroughly. Even if there is no contaminant growth, you can save a day of colonization time in the

The mycelium in this jar has been broken up and is ready to transfer.

future by allowing your mycelium to partially recover from the shake before it's transferred.

A Final Thought

As with most of the methods you encounter in this book, there's no one "best" grain for mushroom cultivation. Consider all the specific factors that affect you before choosing your grain: cost, availability, what substrate you're transferring to, and so on. What's best for some growers may not be best for you.

Grain-to-Grain Transfer

Grain-to-grain transfer is a method of quickly turning a colonized quart of grain spawn into 5, 10, 20, or more quarts of colonized grain. It's a simple method that will save you a lot of time and inoculum once you've mastered the basics of grain spawn.

Using this method requires the use of a glovebox or a flow hood, as the interior contents of the jar will be exposed to the open air. The process involves pouring kernels of colonized grain into a freshly sterilized jar of grain. Growers often use the terms "grain transfer" or "G2G" as shorthand for this method.

The maximum rate of transfer is about one colonized quart (1 L) to 40 uncolonized jars. This is usually written as a spawn rate of 1:40; however, it's best not to attempt a transfer anywhere near this rate. I normally only transfer one colonized quart of grain to 7 to 10 uncolonized quarts. In general, I wouldn't recommend going beyond the 1:10 spawn rate. The further you try to stretch the grain you're transferring, the longer it will take for the transfer jars to finish colonizing. Transferring with a 1:10 spawn rate should yield 10 completely colonized quarts in about 1 week.

One of the key reasons to learn this method is that it saves your inoculum. Buying spore syringes or liquid culture syringes can get quite expensive if you're producing many jars, and you have to wait for them to arrive at your home through the mail. Also, making your own inoculum can be a time-consuming process. You don't want to have to continually make more inoculum when you can simply expand a culture that's already viable.

If you started from a spore solution, a significant part of the colonization time for your jars was devoted to waiting for the spores to germinate. With a grain-to-grain transfer, you'll already have an established culture, so your jar colonization times after the transfer should be much faster than they were when you used the primary inoculum.

A final note: It can be helpful to break up your colonized quart jar 12 to 24 hours before you plan to transfer it. This gives the grain time to recover and will lessen future colonization time. It also allows you to assess the health and vigor of the mycelium prior to transfer. If you break it up and it hasn't recovered the next day, you'll know not to transfer it, as bacterial contamination is likely.

FAQ FOR GRAIN SPAWN

I smell my grain cooking. Is this a problem?
No, it shouldn't be unless you smell something burning. A cooking smell is absolutely normal while pressure cooking grains. A burnt smell means something has gone wrong. You may have forgotten to put your standoff plate in the bottom of the cooker, or your cooker has run out of water.

What kind of water should I use when making jars?
Regular tap water is fine.

My jars look very wet. Is this normal?
Your jars should never look really wet. If it's been a short time after inoculation and the liquid is not from your inoculum, your grains were probably prepared too wet from the beginning. I would let them go and see if you get good growth. If it's been several days since inoculation, the cause could be a bacterial contamination. Smell your filter patch, and if there's a sweet smell, that is the likely cause. You'll need to start over.

What is the purpose of rinsing the grains before loading them into jars?
Rinsing helps get rid of starch buildup on the external shell of the grains. Starch causes the grains to stick together and can hasten bacterial contamination.

Are there any other ways to prevent my grain from clumping?
You can add a pinch of gypsum to the jar. After you load your grain, put in a gram of gypsum and shake it up before pressure cooking.

Is rye grass seed the same thing as rye berries?
No, it is not. Rye grass seed can work as grain spawn for many mushroom species, but it isn't the ideal grain for most species.

My rye grain has sprouted. Will this hurt anything?
That depends on the percentage of seeds that have sprouted and how big they've grown. If only a couple of seeds have sprouted, I would go ahead and use the grain. If much of the rye has sprouted, I would start over. This is usually only a problem if the rye has been sitting around hydrated and uncooked for several days. Pressure cooking will eliminate the viability of the seeds and prevent them from sprouting in the future.

Is wheat an acceptable alternative?
Yes. I would hydrate it by either soaking or pressure cooking it.

FAQ for Grain-to-Grain Transfer

Why do some people use longer or shorter cooking times?
The cooking time depends on the amount of substrate you're cooking. Bigger jars and thicker substrates require more cooking time, so that the heat of the pressure cooker can penetrate to the center of the substrate and sterilize it. Some people sterilize their quart jars of grain for only 60 minutes and are successful. I choose to perform the longer cook. Sterilization is complete once the entire substrate (including the center) has been heated to 250°F (121°C) for 20 minutes.

Is there any way to sterilize jars without a pressure cooker?
Yes. There is a process called Tyndallization, or fractional sterilization (see page 95). I do not recommend this process. It involves boiling the jars on three consecutive days. The results are unreliable, many people have problems with the resulting grain texture, and it's highly inefficient. You can research this process online if you're interested.

How long can my colonized jars of grain sit before I need to do something with them?
It's always best to use your grain quarts as soon after they reach 100-percent colonization as you can. That being said, it's not a problem if they sit for 1 or 2 weeks after full colonization, but don't leave them longer than that. If you anticipate needing to store them for more than a week, put them in the fridge.

Do I have to transfer an entire jar of colonized grain at once?
No, you can save the remainder of your grain spawn for another day if there's some left over in the jar. Just put the lid back on, and it should recolonize into a solid mass just as it was before you broke it up.

How should I incubate jars after a grain transfer?
Nothing special is needed for this process. Incubate them just as you did with your primary inoculum jars.

How many times can I perform the grain transfers with the same culture?
Three times is the generally recognized limit on this. This includes eventually transferring to spawn bags. The more times you transfer, the less aggressive your growth will be, so try to limit this to three transfers. If you started from spores, you could do between five and seven total transfers, but this is not the norm.

Should I shake my jars at 50-percent colonization?
You don't need to. Your jars should colonize quickly enough to make shaking them unnecessary with this process.

Rye Grain Preparation: Soaking Hydration

Three different methods for the preparation of rye grain are described here. The primary difference between these methods is how the grain is hydrated. The first method soaks the grain for 24 hours; the second boils the grain in water for 1 hour; the third pressure cooks the rye and water together in a quart jar, hydrating and sterilizing it in one step. I find 24-hour soaking and pressure cooking hydration to be the two most reliable methods for dealing with rye berries. If you need your rye ready to use in 1 day, go with the pressure cooking hydration method, but all these methods work well, and you should give all of them a try at some point. Remember, what works best in one situation may not work best in all situations. Experiment with many different methods and you'll find the one that works for you.

Materials

- Rye grain, ½ lb. (227 g) per quart jar, plus extra
- Container (to soak rye)
- Strainer
- Inoculum
- Quart jars with lid filters
- Aluminum foil in 8" × 8" (20 × 20 cm) squares, 1 per jar
- Pressure cooker

Time Required

5–10 minutes on first day; about 1 hour on second day

1 **Soak the rye.** Place rye grain into your container. Use approximately ½ pound (227 g) of rye per quart jar you intend to fill, and add a little extra for good measure. Fill the container with water to at least twice the depth of the rye. The rye will expand as it hydrates. Let the rye sit for about 24 hours. Six hours longer or shorter won't be a problem. Add water to the container if the rye becomes exposed to the air.

1

2 **Rinse.** After your rye is hydrated, rinse it off. This removes starch-like material from the exterior surface of the grain, making it less likely to stick together when you shake the jar (see Step #4). Use a kitchen sprayer to rinse the grain through a strainer in the sink. (See page 111 for other rinsing options.) Allow excess water to drain into the sink before loading the jars.

2

3 **Fill the jars and cover.** Fill your quart jars about three quarters full with rye grain. Place a lid with a Poly-fil filter and cover with a piece of foil. Press the foil against the jar, making a foil "cap." Load your cooker with jars and pressure cook for 90 minutes at 15 psi.

3

4 **Inoculate your jars after they are cool.** It will usually take 8 to 12 hours for your cooker to cool completely. Many sources of inoculum can be used, including spore syringes, liquid culture syringes, agar, and grain-to-grain transfers. Shake your jars after they have been inoculated to distribute the inoculum throughout the jar.

4

5

5 **Shake the jars again.** Once your jars have between 25 and 50 percent colonization, shake them again. The goal is to break up the mycelium in the parts of the jar that have been colonized, and distribute those colonized portions into the uncolonized portions. This will speed up the colonization process, saving you several days of colonization time.

After full colonization, you can either spawn the grain to a bulk substrate such as sawdust, straw, or manure; make a casing with the grain; or perform a grain-to-grain transfer to multiply the number of jars you have available.

Rye Grain Variation: Boiling Hydration

Into a pot, pour around ½ lb. (227 g) of rye per quart jar you intend to make, and add a little extra for good measure. Fill the pot with water to at least twice the depth of the rye in the container. The rye will expand as it cooks.

Bring the water to a low boil or a high simmer and cook it for 1 hour. If you cook the rye at too high a temperature, the rye kernels will burst, releasing starches, and the rye will become sticky and more difficult to work with. Grains with burst kernels also seem to have higher rates of contamination than those without.

Continue from Step #4 of the soaking hydration method.

Rye Grain Variation: Pressure Cooking Hydration

Scoop 1 cup (227 g) of rye grain into each jar you intend to make. Measure ⅔ cup (160 mL) of water and pour it into each of the jars.

Place the lid (with filter) on the jar. Place a piece of foil over the jar. Press the foil against the jar, making a foil "cap." Load your cooker with jars and pressure cook for 90 minutes at 15 psi. Continue from Step #4 of the soaking hydration method.

Popcorn: Pressure Cooking Hydration

Materials

- Popcorn, ½ lb. (227 g) per quart jar, plus extra
- Quart jars with lid filters
- Water
- Aluminum foil in 8" × 8" (20 × 20 cm) squares, 1 per jar
- Pressure cooker
- Strainers
- Inoculum

Time required

4–5 hours

1 **Fill the cooker with corn.** Remove the standoff plate from the bottom of your pressure cooker. The cooker must be completely empty. Place ½ lb. (227 g) of popcorn into the bottom of the cooker for each jar you want to make. Fill the cooker with dry corn no more than halfway up the side.

2 **Pressure cook the corn.** Fill the cooker with water. The water level should be near the top of the cooker and at least twice as deep as the level of corn in the cooker. Pressure cook the corn at 15 psi for 1 hour. After the corn has cooked and all the pressure has been released, move the cooker near the sink while it's still hot. Be sure to put a towel underneath the cooker so it does not scratch or burn the counter or floor.

3 **Rinse the popcorn.** Scoop the corn out of the cooker with a strainer or large bowl and into strainers over the kitchen sink. *Be careful not to burn yourself with the water.* Wash the corn off using the kitchen sprayer. Removing and washing the corn while the water in the cooker is still hot leaves most of the starch, in the water. If the cooker cools down with the corn in it, the water will turn into a solid mass of starch, and the corn will be ruined. Allow the corn to drain in the strainers.

3

4

6

4 **Fill your jars.** Load your quart jars two-thirds full with the washed corn. Place the lid, with filter, on the jar. Place a piece of foil over the jar and press it firmly against the jar, making a cap.

5 **Pressure cook the filled jars.** Rinse out your pressure cooker with clean water before proceeding. This will remove any extra starch left on the sides of the unit. Replace the standoff plate, load your cooker with the jars, and pressure cook them for 60 minutes at 15 psi. Let the jars cool completely in the cooker.

6 **Inoculate your jars after they're cool.** Don't open your cooker until you are ready for inoculation. Many sources of inoculum can be used, including spore syringes, liquid culture syringes, agar, and grain-to-grain transfers. Shaking your jars to speed colonization is usually not necessary for popcorn; however, if If one side of the jar is not colonizing, it may be necessary.

You have several options after full colonization: you can spawn the grain to a bulk substrate such as sawdust, straw, or manure; make a casing with the grain; or perform a grain-to-grain transfer to multiply the number of jars you have available.

Wild Birdseed/Millet: Soaking Hydration

Materials

- Millet (or wild birdseed with millet as main ingredient), ½ lb. (227 g) per quart jar, plus extra
- Container for soaking grain
- Quart jars with lid filters
- Aluminum foil in 8" × 8" (20 × 20 cm) squares, 1 per jar
- Pressure cooker
- Inoculum

Time Required

5–10 minutes on first day; about 1 hour on second day

1 **Soak the millet.** Place millet into your soaking container. Use about ½ lb. (227 g) of millet per quart jar, plus a little extra for good measure. Fill the container with water to at least twice the depth of the millet. The grain will expand as it hydrates. Let the millet sit for about 24 hours. Six hours longer or shorter will not be a problem.

2 **Rinse your millet after it's hydrated.** This removes starch-like material from the exterior surface of the grain, making it easier to shake the grain in the jar. I usually use the kitchen sprayer to rinse the grain in a strainer over the sink. Other rinsing options are suggested on page 111. Let the strainer sit for 5 to 10 minutes to allow excess water to drain off the grain into the sink before loading the jars. Fill your quart jars about three-quarters full with washed millet.

3 **Place the lid, with filter, on the jar.** Place a piece of foil over the jar. Press the foil against the jar, making a "cap."

4 **Pressure cook the filled jars.** Load your cooker with jars and pressure cook for 90 minutes at 15 psi. Once the cooker pressure is down to zero and the cooker has cooled for an hour or two, remove the jars in your work area and shake them. If you don't shake them while the jars are still warm, the grain can solidify in the jar, making it impossible to shake.

5 **Inoculate your jars after they have completely cooled.** It usually takes between 8 and 12 hours for a pressure cooker to cool down completely to room temperature. Many sources of inoculum can be used, including spore syringes, liquid culture syringes, agar, and grain-to-grain transfers. Shake your jars after they have been inoculated to distribute the inoculum throughout the jar.

6 **Shake the jars.** Once your jars reach 25- to 50-percent colonization, shake them again. The goal is to break up the mycelium in the parts of the jar that have been colonized, and distribute those colonized portions into the uncolonized portions. This will speed up the colonization process, saving you several days of colonization time.

You have several options after full colonization. You can either spawn the grain to a bulk substrate such as sawdust, straw, or manure; make a casing with the grain; or perform a grain-to-grain transfer to multiply the number of jars you have available.

Grain-to-Grain Transfer

Prepare quarts of your favorite grain using one of the methods described in this chapter. It's best to fill the jars a little less full with grain to allow extra room for the colonized grains you'll be transferring into the jar. Pressure cook these freshly prepared grains. Allow them to cool completely.

1 **Break up the grain.** Break up the quarts of the colonized grain you'll be transferring. Review the options for breaking up jars (see page 113). Place these colonized quarts and your freshly prepared grains into your glovebox or in front of your flow hood. I suggest wiping off the colonized quarts with isopropyl alcohol or spraying the exteriors with some disinfectant. They've been sitting out in an unsterile environment during colonization, and could have picked up some contaminants on the external surface that you don't want dropping into your freshly sterilized grains. This just eliminates one additional possible source of contamination.

2 **Transfer the grain.** Open a colonized jar and one of your uncolonized jars. Pour some of the colonized contents into the uncolonized jar. This will usually be the equivalent of 2 or 3 spoonfuls of spawn to get near a 1:10 transfer rate. Some people sterilize a spoon and scoop the spawn in, but I find this to be an unnecessary step that can potentially transfer contaminants between jars.

3 **Shake and incubate.** Close the mixed jar and set it aside. Repeat the process of opening the uncolonized jars and pouring in the sterilized substrate for your remaining jars. After you've finished your transfers, shake up your jars thoroughly to distribute the spawn evenly in the jar. The more evenly it's distributed, the faster your jar will attain 100-percent colonization. Place your jars in their incubation location.

CHAPTER 7

CASINGS

Once you can reliably generate grain spawn in quart jars, the more advanced processes in the mushroom cultivation world open up to you. At the end of the last chapter, you learned about grain-to-grain transfer, the first step most growers take when they have colonized quarts. Expanding the mycelial mass is always your goal in cultivation, and the ease with which the grain-to-grain method can be accomplished makes it well worth the time invested. It turns your 5-quart jars of colonized grain into 35 to 50 quarts in less than 2 weeks.

But what should you do with these freshly colonized quarts? Growers of some species go directly to fruiting from the grains. To fruit, some species require that the grains be broken up into the bottom of a pan and covered with a *casing*, a layer of non-nutritious material that helps retain moisture. *Agaricus* and *Psilocybe* species most commonly use casings. If you're primarily interested in wood decomposers, such as Shiitake and Oyster mushrooms, this chapter will not apply to you.

Casing Considerations

The easiest way to fruit directly from your grains is to make a simple casing, a layer of material that is applied over the top of your grain spawn or bulk substrate to keep the substrate underneath from drying out and provide your fruiting culture with enough water to sustain fruitbody development. Common materials used for casing layers are vermiculite, coconut husk fiber (coir), and peat moss. Methods for making casings from all these materials are included in this chapter.

The two main components of a casing are the substrate and the casing layer. After the substrate is combined with its casing, I will often refer to the combined substrate/casing as simply "the casing" or "casings."

Preparing *basic* casings is a three-part process: you pasteurize your casing material, hydrate it, and apply it to the top of your substrate. Your substrate can either be fruited directly out of the top of the jar, or broken up and poured into the bottom of a pan. I often refer to these as "straight-grain casings" or "simple casings" because they don't make use of a bulk substrate; however, not all species can be fruited this way, because many require spawning to a bulk substrate before the casing layer is applied.

Pasteurization

Pasteurization is different from sterilization. The goal of sterilization, to kill off living organisms, is achieved at temperatures above 250°F (121°C) for 20 minutes. Pasteurization is performed at much lower temperatures, which allows beneficial microorganisms to survive. Many species of mushrooms, particularly *Agaricus* species, begin fruiting faster and offer better yields when these microorganisms are present in the casing layer. Pasteurization is usually performed in a hot water bath of 140 to 160°F (60 to 71°C) for 1 hour. Although sterilization is the ideal heat treatment for the preparation of grain substrates, pasteurization is the ideal heat treatment for preparation of casing materials.

Size

The size of your casing is determined by the size of the container you use for fruiting. Almost any type of container will work. Fruiting containers can range in size from a coffee mug to a wooden structure the length of a large growhouse. At-home growers often use plastic food-storage containers, pie tins, aluminum pans, glass bowls, storage bins, etc.

There are several things to consider when determining casing size. You don't want to make your casing too large in case you have to throw the entire thing out should any contamination arise. While it's sometimes possible to save a contaminated casing, you'll often have to throw it away lest the contamination spread to other casings. This means that much of the substrate you spent weeks crafting may be lost at one time. On the other hand, making a casing too small may limit the fruiting ability of the substrate and your final yields.

The smallest size I would recommend for any type of casing would be an 8" × 8" (20 × 20 cm) pan. Thin aluminum pans are inexpensive and sold in big box and grocery stores. These square pans are the perfect size for 1 quart of grain or three or four PF cakes. The largest size I'd recommend would be about the size of a turkey roasting pan. Large turkey roasting pans require about 4 quarts of colonized grain as the base substrate. Plastic storage bins of this size also work well. Casings can surely be made larger than this, but you shouldn't consider it until you have a significant amount of experience.

Casing material can be placed in canning jars for pasteurization.

Be sure your container doesn't have any holes in the bottom. You'll usually get multiple harvests from your casing, and between harvests you may want to submerge and rehydrate your casing by filling the container with water. This will improve your yields for the next flush of mushrooms.

Substrate Thickness

In general, the thicker the substrate layer beneath a casing, the larger the fruits will be and the better the yields. But your substrate is a precious commodity. Making a thick substrate layer means that you're using a large amount of substrate. Thinner substrate layers allow you to make more total casings with a greater surface area for fruiting. This is the approach I recommend.

Commercial growers who are using bulk substrates such as compost will often make substrate layers that are 4 inches (10 cm) deep or more. This is not necessary for home growers, especially if you're fruiting directly from grains. If you're making straight-grain casings in an 8" × 8" (20 × 20 cm) container, a substrate thickness of ½ to 1 inch (1.3 to 2.5 cm) is the ideal balance between thickness and surface area. Fruiting abilities at each thickness are likely to be dependent on the genetics of the species. It's always best to try several different substrate depths to see which works best for you.

Casing Thickness

The thickness of your casing layer depends on the thickness of your substrate, the casing material you choose, and the species you're growing. Remember that one of the key goals of a casing layer is water retention. Using a casing material such as vermiculite means that you can use a thinner layer than with some other materials.

Larger casings will generally be made with a greater substrate depth, and a thicker casing layer. If you're using a bulk substrate like manure, you'll also want a thicker casing layer. For most home-scale grows, you will never need a casing layer over 1 inch (2.5 cm) thick, even if you're using a bulk substrate. For most straight-grain casings, you should not need a casing layer over ½ inch (1.3 cm) thick.

Casing Method

This process is demonstrated with vermiculite, but peat or coir may also be used as casing layer material. See page 133 for preparation guidelines for each of these materials.

Materials

- Colonized substrate (quarts or PF cakes)
- Fruiting container
- Vermiculite
- Large microwave-safe bowl
- Rubber gloves
- Isopropyl alcohol
- Tap water

Time Required

3–5 minutes per casing

1 **Microwave the vermiculite.** Fill your bowl with vermiculite. Microwave it dry for 5 minutes. Hydrate the vermiculite by filling the bowl with cool water from the faucet or kitchen sprayer. Put on your rubber gloves. Wipe them down with alcohol.

2 **Prepare your substrate.** If you're using quarts, your jars should be broken up. If you're fruiting from a larger container, wipe down the interior of your fruiting container with alcohol. Let it dry and pour your grains into the bottom of the container. Pack your grain substrate down gently. The entire surface of the substrate should be level. Also consider the Reservoir Effect (see page 132).

3

3 **Squeeze the vermiculite.** Using both hands, remove some vermiculite from the water and squeeze it gently. Give it one quick and semi-firm press. That should get the water content right; water shouldn't be dripping from the lump of vermiculite.

4a

4 **Crumble the ball of vermiculite.** Spread it over the surface of the substrate. Do not compress or pack down the vermiculite layer, although you may have to rub your hand back and forth over the surface to spread it out. The casing layer should be as smooth and level as possible, but not packed down. The smoother you're able to get the substrate and casing layers, the more even your harvest will be. Start your vegetative growth cycle or put your container into a fruiting chamber.

4b

Casing Materials

Although online and print mushroom cultivation resources suggest elaborate components and processes for building the "ideal" casing material, the goals of small-scale growers are somewhat different from those of most commercial growers and scientists. As a home grower you can choose from among several simple, accessible materials to make your casings.

Vermiculite

The best option for home cultivators using straight-grain casings is often a plain vermiculite casing layer. Used in gardening to help retain water, vermiculite performs particularly well as a casing layer. It's readily available and can be prepared in minutes. It's also very low in nutrients, making contaminants less of a concern. Vermiculite granules range from nearly marble-sized to the size of grains of sand. The ideal size for making vermiculite casings is a little smaller than the size of a pea, and is usually labeled as "coarse" on the bags. As with most things in mushroom cultivation, grain sizes a little larger or smaller won't have much of an effect on the end result.

Vermiculite is often ready to use as a casing layer directly out of the bag, with no further heat treatments required, as its manufacturing process leaves it sterile enough to work with. Making casings is not considered to be a sterile procedure — you make the casings and fruit them in an open-air environment. Even though the strict sterile procedures that governed many of the methods up to this point are no longer required, I still recommend sterilizing vermiculite prior to use. Unlike other casing layers, such as peat moss, it doesn't contain many microorganisms that are beneficial to the mushroom colony. Thus, further sterilizing the casing layer will not have harmful effects.

One of the simplest ways I have found to prepare vermiculite is in the microwave. Microwave dry vermiculite in a heavy-duty bowl for 5 minutes. The vermiculite will quickly heat up to several hundred degrees Fahrenheit in this short time, and any potential contaminants will be killed. Be sure to use a heavy-duty plastic or glass bowl for this process, as most common plastic bowls will melt during cooking.

Hydrate the vermiculite after you pull it out of the microwave; this will cool it down to a safe temperature, both for your hands and for the mycelial culture. Tap water is fine for hydrating the casing layer.

Reservoir Effect

This is another supplemental method that you should consider if you plan on casing straight grains. It's simply the addition of moist vermiculite to your substrate layer before you add the casing layer. This method has two benefits. First, it increases the moisture reserves that are available to your substrate as it regenerates into a solid mycelial mass, giving it an additional "reservoir" of water. Second, it increases the volume of your substrate layer, allowing you to make more or larger casings with the substrate you have available.

Think of this process as an addition to step 2 of the vermiculite casing method described on page 131. Add vermiculite to your substrate layer in a 3:1 substrate-to-vermiculite ratio. If you're putting three quart jars of spawn in your fruiting container, then add one additional quart jar of moist vermiculite to the pan as well. Mix the vermiculite up with your grain, and even it out.

Some growers add a thin layer of moist vermiculite underneath the grain substrate layer in their casings instead of, or in conjunction with, mixing additional vermiculite with the substrate. Although I don't see much additional benefit over the mixed vermiculite reservoir effect described above, you can experiment with it and see how it works for you.

Adding vermiculite to bulk substrates, such as compost and manure, is often beneficial for the same reasons as those described here. These processes are described in greater detail in chapter 13.

Other Casing Materials

Vermiculite is easy to work with and readily available, but you can use other materials for casings as well. These include coconut husk fiber (also called coir) and peat moss. If you try some of these other materials, it's best to mix them with vermiculite because of vermiculite's exceptional ability to retain water.

Coconut Husk Fiber (Coir)

Coir, coconut husk fiber, coconut grow medium, coir bricks, and coconut fiber substrate are all names for the same product. Derived from the brown fibrous material on the outer shell of coconuts, coir is often sold compacted into bricks. Once you open the package, the compressed brick must be soaked in water for 10 to 15 minutes, where it will expand to about five times its original volume. The result is

Coir is commonly sold in compressed bricks. Before it can be used, it must be soaked in water.

a soft, spongy material that is fairly moisture retentive.

Casing layers are exposed to semi-sterile air from the time they're made to the time they're disposed of, so any casing layer should be non-nutritious to prevent easy contamination. Although some people argue that coir contains more nutritional content than vermiculite, I haven't experienced contamination problems with it. It contains roughly the same nutritional content as peat moss, the most common casing material in the mushroom industry. If contamination is a concern, add a bit of hydrated lime or another form of calcium carbonate to the casing layer to buffer its acidity. A higher pH will make the casing less susceptible to invading contaminants.

One of the best things about including coir in your casing mixture is that it tends to change color as moisture levels change. As it dries out it becomes a lighter brown, so it's great to include it in the casing layer mixture as a natural moisture monitor.

Coir is sold in gardening and hydroponic stores as a component of soil for plants. It can also be found at pet stores, where it's sold as reptile bedding. It's generally sold in a package a little smaller than a common construction brick.

Using Coir as a Casing Material

Prepare coir casings in much the same way as you prepare vermiculite casings. Most people use a 50:50 mixture of coir to vermiculite. After your coir brick has expanded, wring out all the extra water, and mix in an equal amount of vermiculite. Microwave the mixture for 5 minutes, and add water to cool and hydrate it.

Many growers who use coir in their casing material mix it with vermiculite in a 50:50 ratio.

Alternative Pasteurization Methods for Coir

Another common way to prepare this casing material for use is to pasteurize it. Fill quart canning jars with your hydrated casing material and boil them for an hour on the stove. If the casing material's moisture content is right before you pasteurize it, you can simply pour it out of the jar and onto the substrate once it cools down. Larger amounts of casing material can be pasteurized in the oven using oven bags, available at most supermarkets. Spawn bags can also be used for pasteurization, but they must be boiled in water on the stove, not in the oven. They're not designed to withstand dry heat.

Coir as a Bulk Substrate

Most species of mushrooms will colonize and fruit from a large variety of potential substrates, whether they have substantial nutritional value. If you want to try coir as a bulk substrate component, I'd encourage you to give it a whirl. It will surely colonize and fruit when used in that manner as well.

Peat Moss

Peat moss, or sphagnum, is the most commonly used casing in the mushroom industry. Most large-scale mushroom farms use it as the primary ingredient of their casing layer. Peat moss comes from peat bogs, wetland areas that accumulate dead organic material over long periods of time. Decomposition of the material in these areas by bacteria and fungi is slowed because it's continually inundated with water and most of the oxygen in the system has been removed. Peat moss also contains beneficial microorganisms, so if you're attempting to grow *Agaricus* species such as White Buttons, this is the casing material to use.

Peat moss is readily available in spring and summer at lawn and garden centers. In winter, when many smaller garden centers are closed, you should be able to find it at big box hardware stores.

Using Peat Moss as a Casing Material

Because of the way it's formed, peat moss is much more acidic than the other casing options presented in this chapter, so you'll need to thoroughly mix in 3 tablespoons of lime (calcium carbonate) and 3 tablespoons (43 g) of gypsum (calcium sulfate) per quart of casing material (so, 3–5% of each by volume) to reduce the acidity. You're aiming for a 1:20 ratio of gypsum and lime to peat by volume, creating an ultimate pH of from 7.5 to 8.5. Amounts may vary, depending on the acidity of your brand of peat moss. Peat moss can be quite fine, so seek a brand with a fibrous texture; mixing it 50:50 with vermiculite helps improve the texture of the casing layer.

This mixture of peat, vermiculite, lime, and gypsum should be moistened and pasteurized before it's used as casing layer material. To judge the correct moisture level after adding water, pick up the casing. No water should drip from the casing layer material unless it's lightly squeezed. If you squeeze the material and no water comes out, add a bit more water. Fill quart canning jars with the moistened and mixed casing material, apply lids, and boil for 1.5 hours on the stove. Let the mixture cool before using.

Once you've made your casings, you have a choice to make: start your casing in a vegetative cycle, or place it directly into the fruiting chamber. If you decide to transfer directly into a fruiting chamber, you can choose one of the setups from the PF Tek discussion, using perlite to raise relative humidity (see page 72) or one of the more advanced fruiting chamber setups (see chapter 10). Deciding which setup to use will most likely depend on how many casings you have made.

Entering the Vegetative Cycle

When you first make your casing, the substrate layer will still consist of the individual kernels of grains that were visible after you broke up the quart jar. The casing will need to enter a period of vegetative growth to be able to recover. This phase generally requires a period of low light, higher CO_2, and higher temperatures. A common way that some growers achieve this is to cover their casings with foil for two or three days once they're made. This provides darkness and will keep the CO_2 at a higher level. After 2 or 3 days, the individual kernels should form back into a solid mycelial mass in the bottom of your casing pan. This should give the casing a bit of form and structure. If your casing does not solidify after the third day, you may have a contamination issue (see page 45).

On a small scale, it can be a good idea to cover straight-grain casings with foil and let them recover for several days before introducing them to the fruiting chamber. This isn't absolutely essential, but on a larger scale, when using bulk substrates such as compost or manure, it usually makes sense to use this vegetative period. Experiment with it both ways.

Mycelium should begin to break through to the surface of your casing layer shortly after the substrate layer fully solidifies. The amount of growth you see at the surface will depend primarily on how thick the casing is, how evenly it was applied, and the vigor of the species you're growing. You don't need to have any visible white mycelial growth at the surface of the casing for mushrooms to form, although most healthy casings will have some mycelium breaking through.

End this additional vegetative cycle and initiate pinning once your casing has solidified again, and you start to see some mycelium breaking through the surface layer.

Pinhead Formation

Once your casing enters the pinning phase, the only two things you need to do, at least for the first week, are to maintain the humidity and provide fresh air for your fruiting chamber. The goal is to maintain a relative humidity above 90 percent during this time. (See chapter 10 for tips on getting humidity levels right in midsized fruiting chambers.) To ensure that there's adequate air exchange, fanning out your grow area should do the trick, as directed in the section on PF fruiting chambers (see page 56).

Mycelium grows during spawn and vegetative phases with high levels of CO_2. This is primarily because mycelium respires CO_2 as it grows, and the gas has nowhere to escape in the sealed containers. Once you put your casing into the fruiting chamber and fan it often, you'll drastically reduce the levels of CO_2. This is one of the triggers that tell the culture it is time to begin fruiting. A second trigger that you'll be using is to expose your casing to light. As with the PF Tek, a cycle of 12 hours light/12 hours dark works well with most species.

There's one other thing to think about before your mushrooms begin to form. One week after the casing is made, you should give your casings a good spray down with a water bottle or sprayer. At about a week and a half, you should start seeing the first young mushrooms (also called *primordia* or *pins*) starting to form on the surface of the casing layer. Moistening the casing before this occurs will ensure that no damage is done to the pinset as it is forming.

Be careful never to spray water directly onto the primordia because it could inhibit their growth. Most sprayers forcibly discharge pressurized water, and the force of impact can damage pins that are forming. The best way to spray is to get the mist as fine as possible and spray at least a foot above the surface of the casing. This reduces the amount of force with which each droplet falls. Be sure to write down the date you made your casings and hydrate them at the one-week mark, as you don't want to interfere with the primordia formation.

Keep in mind that these young mushrooms don't do well with any other significant disturbances at this critical time. Other types of disturbance that could be damaging to your pins include significant changes in temperature or humidity, or the introduction of chemicals of any kind. I once killed off an entire harvest of mushrooms by wiping down the interior surface of the small grow chamber with a strong bleach solution during pinhead formation. The next day, every mushroom in the container had died off.

With every casing, a small percentage of the mushrooms that begin growing as pins will abort. This percentage will vary depending on species, strain, and environmental conditions. If a very high number of your pins are aborting, you may be overlooking a problem of some kind. If only a small percentage of mushrooms are aborting, this is probably a normal reaction to the indoor cultivation processes.

The substrate begins to "pin" or form mushrooms.

Growth Problems

It's important for inexperienced growers to watch their young mushrooms carefully. The following two problems can be solved if caught early enough.

Overlay

While surface growth is generally a sign of health and vigor, sometimes the mycelium begins to cover the surface of the casing layer, and the casing colonizes unevenly. This is often caused by applying the casing layer unevenly. Ideally, your casing layer should colonize evenly, as this will help create a more even pinset and ultimately a better yield.

Overlay, or uneven growth of the mycelium, can cause a drop in yield because of decreased pinning. The solution is "patching" — sprinkling a thin layer of casing material on the areas that exhibit overlay.

Solution: Patching. Some growers choose to "patch" casings that show signs of overlay. This simply involves applying an additional thin layer of casing material to the areas that show signs of overlay. This process will allow the uncolonized areas to catch up in growth, and should ultimately produce an evenly colonized casing.

Matting

On occasion, an area of overlay will appear compacted and hard, and may actually become a little hard to the touch. This is referred to as being "matted," and is detrimental to your culture. A matted casing is usually the result of overwatering. Watering compresses the overlay and produces a hard surface. Ironically, even though this growth is produced by overwatering, if left alone it will prevent water from being absorbed down into the casing, and will ultimately inhibit the formation of primordia wherever it occurs. Another possible cause of matting is that your casing was too dry. Mushroom mycelium can take on several different growth forms depending on different environmental stimuli. If your casing is too dry, a type of mycelia can form that's very similar in appearance to matted overlay. This will also form a hard surface on the casing and inhibit water from penetrating the casing layer.

Solution: Scratching. The solution to a matted casing is a process called "scratching." As the name suggests, the best option is to use a fork from your kitchen to scratch the matted surface. First, wipe down a kitchen fork with rubbing alcohol to sterilize it. Then literally scratch the surface of the casing to break up the parts that have a matted appearance. It's advisable to scratch the casing only as a last resort, as it will make it easier for contaminants to set in the disturbed area. Scratching is also a good solution for a casing that just doesn't seem to want to fruit, even though the conditions seem right. If it's been several weeks and no pins have formed, scratch the surface and put the casing back into the fruiting chamber.

Scratching the surface of a casing with a sterilized fork will break up matted casings.

Contamination

Once your casings are in the fruiting chamber, your primary concerns are maintaining humidity levels and providing an ample supply of fresh air. Assuming you have those two variables under control, there's only one other thing you need to watch out for: contamination. If your casings become contaminated by molds, your months of hard work can be lost. Contamination is the most common cause of failure in mushroom growing.

Fortunately, it's not necessary to throw out an entire casing at the first sign of contamination. There are several methods you can use to try to save the casing. These methods may not always be successful, but every grower should be aware of them.

The two most common types of contamination to watch out for in casings are green mold and cobweb mold. These are described more thoroughly in chapter 3, but it may be helpful here to provide a brief overview of what you may encounter at this point and some solutions for dealing with these common types of contaminants.

Cobweb Mold

The most common contaminant of the casing layer for the home cultivator is cobweb mold. It can sometimes resemble the mycelium you're trying to grow, but is generally much wispier in texture and often develops some gray tones. It usually starts as a small white patch on the casing, but grows very quickly, sometimes colonizing the entire casing layer in a matter of days. For this reason, it needs to be identified and controlled as soon as possible.

There are several ways to identify cobweb mold soon after the first spots appear. First, its texture is different from that of mycelium, but it's often hard for the new grower to tell the difference. The growth of cobweb mold is usually much thinner and finer than that of most types of mushroom mycelium. It truly resembles thin strands of a cobweb, while mushroom mycelia are usually thicker and more robust. Cobweb mold also often begins

Cobweb mold, a common contaminant, can colonize a casing in a matter of days.

to rise visibly above the casing layer as it grows upward; however, the key identifier is speed of growth. If you see a thin, wispy, grayish spot that has grown significantly in 6 or 12 hours, there's a good chance it's cobweb mold.

A 3-percent solution of hydrogen peroxide, available in groceries and pharmacies everywhere, is an effective tool for battling a cobweb mold contaminant. To use this solution, screw the spray nozzle from a similar bottle (not from a bottle of household cleaner or anything that may damage your mushroom culture) onto the peroxide bottle. Spray the area of contamination several times until the spot is covered with peroxide. Repeat this process once or twice a day for the next couple of days.

Also increase the fresh air and slightly lower the humidity in the fruiting chamber until the mold has stopped spreading. If you see the spot continue to grow, you'll have to take additional measures, such as trying to cut it out and "salt the wound" (see page 141).

Green Mold

A second fairly common type of contaminant is green mold. It's generally a species of *Trichoderma*. When this contaminant first arises, it may also look like mushroom mycelium, but it will appear as a solid mass of white, rather than individual hyphae stretching out. Shortly after the white mass forms on the surface, it will start to turn green. When you see the green forming, it means that the spores of the mold are being released. Any casing with green mold on it should be immediately removed from your growing area. This type of mold is also very aggressive and fast growing. Because it begins to produce spores soon after it appears, the spores can spread throughout your growing area, making future contamination more likely, which is why green mold is seen as one of the most detrimental contaminants in the mushroom world.

Salting Wounds

Hydrogen peroxide is not very effective against green mold, and spraying it will just blow the spores around your growing environment. The best thing you can do to save your casing from green mold is to cut away the contaminated area. Sterilize a spoon from your kitchen with alcohol and use it to scoop out the contaminated portion of the casing into a separate bowl. Take a little extra around the edge of the contamination to be sure that you're removing the entire colony. After it has been removed, pour some common household table salt onto the uncontaminated casing around the area you just removed. This will prevent any further growth of the contaminant.

It's entirely possible that one or more areas of contamination may arise after you've taken these measures. If this happens, you may have to throw away the entire casing. Don't keep a highly contaminated casing in your growing area for too long, because the spores that the contaminant produces may end up causing you problems in future attempts at growing. Every grower has to throw away a lot of hard work because of contamination at some point. The worst thing you can do as a grower is to contaminate your home with high levels of mold spores, as this will give you headaches long into the future.

You can save casings with just a small amount of contamination by removing the infected area and "salting the wound."

Fruiting and Harvest

It should take 1.5 to 2 weeks for your mushrooms to begin to form. Depending on the species, it will often take 4 to 5 days for your primordia to form fully into mushrooms. During this time, maintain the conditions you established during the pinhead initiation cycle. Other reference books for mushroom cultivation will recommend different temperature and humidity levels for pinhead initiation and fruitbody development. As a small home grower, you may not possess the required equipment to consistently monitor or alter the humidity by 10 percent or to raise the temperature by 5 degrees. But don't worry: just maintaining consistent conditions in your fruiting chamber will suffice.

If your mushrooms abort or start to grow in some deformed way at this point, there could be many different factors involved. One of the most common is high CO_2. Inadequate ventilation of your growing environment will cause mushrooms to form skinny stems and small caps. It will also significantly depress the yield of the resulting harvest. Increasing the amount of ventilation should solve this problem with the next harvest.

It's best to wear gloves when harvesting mushrooms, as they prevent potential contaminants on your hands from contacting the surface of the casing and affecting future harvests. They also keep spores from getting all over your hands.

To harvest, simply grasp your mushroom or cluster of mushrooms near the

These *Agaricus* species mushrooms are fruiting from a casing. Once they're harvested, the casing can be rehydrated to encourage further fruiting.

base and gently twist it off the casing layer. It should come off fairly easily. This is also the best time to remove any casing material from the base of the mushroom. If you leave a lot of the casing layer on at this time, it will get mixed into the rest of the mushrooms in your collecting basket and will be difficult, if not impossible, to remove later. Many growers of edible species simply cut the base off each mushroom before placing it in their collecting bin.

When harvesting mushrooms from a casing, your goal should be to disturb your casing as little as possible. If a large chunk of substrate comes up with the mushroom you just harvested, as it often does, it's fine to go ahead and place that chunk back into the casing container where it was. Sometimes, broken pieces will reattach to the rest of the main substrate layer. Even if they don't reattach, a mushroom or two can still form from the piece that has broken off if you decide to save it.

Dunking after Harvest

Dunking is the term used for submerging a substrate in water to rehydrate it between harvests. Normal dunk times are between 12 and 24 hours. To dunk a casing, one good option is to fill a gallon jug with tap water. Next, simply pour the water onto the casing layer of the substrate. This is likely to displace some of the casing material around the top and off the sides, but this is normal and shouldn't be a concern. Now let the casing sit on a shelf, inside or outside the grow area, for 12 to 24 hours. After the recommended time has elapsed, pour the water out into a separate bowl or bucket. You'll probably lose some of the casing layer that was displaced, but once again, that shouldn't be a problem. After you pour out the water, put your casing back into the grow area and resume fruiting conditions. The next flush should come in about a week.

Dunking Tips

When dunking, it's best to wear gloves and hold down the casing with your hand as you pour out the water. Even if you pour carefully, the entire casing may slip out of the pan and into the bucket if you don't hold it back. Also, don't pour the dunk water down your sink; it may contain bits of casing and chunks of the substrate layer. Instead, pour it into the toilet, a bucket, or outside into a compost pile.

FAQ for Casing

Are there any types of lime to avoid in my casing layer?

Yes, any variety of lime with more than 5 percent magnesium should be avoided. Dolomitic lime is one example. Also, don't use blackboard chalk.

The mycelium seems to have eaten through my aluminum pie pan. Is this normal?

Yes. Mushrooms produce a huge number of digestive enzymes that are capable of breaking down most materials, some more quickly than others.

CHAPTER 8

LIQUID CULTURES

Liquid cultures are used in many mushroom-growing operations as a way to propagate a culture in liquid form. This is very valuable for the home cultivator because it can turn one spore syringe or petri dish into dozens of syringes that can be injected into jars. A single spore syringe may make five or six grain jars, but a spore syringe inoculated into a liquid culture can yield enough fluid to inject hundreds of grain jars. This type of mycelial suspension in liquid is a very valuable asset for a cultivator.

A liquid culture consists of a nutritious material that is mixed with water and then sterilized. This nutritious material is commonly barley malt sugar, also known as maltose, malt extract, light malt sugar, or light malt extract. This can be found at brewing or health food stores. Another common nutritional material used in a liquid culture is light corn syrup such as Karo. This liquid solution is injected with spores or another living culture, and allowed to incubate. After a week or two of colonization, the result is a large amount of mycelial mass suspended in the liquid. This is a liquid culture.

Learning to perfect liquid cultures means that you won't need to buy or create nearly as many spore syringes or agar plates for inoculations. It saves you a significant amount of time and money in the long term and is a process worth mastering.

Scalpels can be sterilized in a pressure cooker or with an infrared sterilizer such as the one shown here.

Sterile Procedures Are Vital

At this point, it's important to note that the procedures in this chapter are much more prone to contamination than many of the other methods described, and any contamination during these procedures will have negative consequences for every step that follows. For this reason, you must take extra care to ensure that you have the proper clean procedures in place, that you have a pressure cooker for proper sterilization, and that you use a flow hood or a glovebox.

Preparing a Liquid Culture

1 **Fill your canning jar two-thirds full of water.** Add 1 tsp* of light malt sugar or light corn syrup per 100 mL (3.4 ounces) of water.

2 **Add a marble or nickel.** This will help break up the mycelium later.

3 **Prepare the jar and pressure cook.** For pints and quarts, use lid with a Poly-fil filter on the jar of your liquid culture (see page 103). For half-pint jars, use a punctured lid, the punctures covered with masking tape (see page 58). Screw on the ring band. Cover the jar with foil and pressure cook for 15–20 minutes at 15 psi.

4 **Inoculate the jars.** After allowing the jars to cool, inject 1 or 2 cc of spore fluid per half-pint jar. To keep the materials sterile, perform the inoculation in a glovebox or in front of a flow hood. Place the inoculated jar in a warm, dark place and shake the jar twice daily. It will be fully colonized after 2 weeks. Once the jar is fully colonized, store it in the refrigerator.

1

2

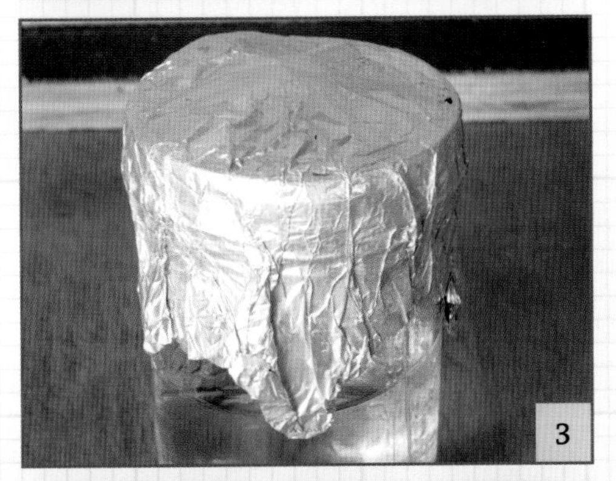

3

4

*Light malt sugar: 1 teaspoon = 3g
Light corn syrup: 1 teaspoon = 5g

Pressure Cooking Liquid Culture

When pressure cooking any liquid culture, the main thing you need to watch out for is caramelization of the sugars. If this occurs, it can cause abnormal growth of the culture. Sugars become caramelized when they're exposed to high temperatures for too long. You can usually see this easily: caramelized liquid appears more amber when it comes out than when it went in, and there are usually floating remnants of sugars that have bonded together or debris collecting at the bottom. Ideally, the liquid should be the same color after sterilization as it was before. Different substances can withstand different temperatures, so if you're having problems, try bringing the pressure cooker up to temperature more slowly than you normally would. If you're still having problems, try lowering the pressure to 5 or 10 psi and cooking the mixture for 30 minutes instead of 15.

Inoculation

There isn't really a special method to use when inoculating liquid cultures. It's the same as any other inoculation procedure. The most important aspect is sterility, so be sure to use a flow hood or glovebox, use gloves, and plan your movements in advance. As always, move with confidence and precision. For half-pint canning jars, inoculate with the same amount of fluid as you would a PF jar, which is about 1 or 2 cc of spore fluid or liquid culture. For larger liquid cultures, use more.

The inoculated liquid culture should be placed in a warm, dark place for colonization. Shake up or swirl your liquid culture twice a day, every day. This will help to get oxygen down into the culture. The marble or nickel will also break up and distribute the growing culture around the jar. More advanced methods use a magnetic stir bar and stir plate to aerate the culture.

Storage

It can be difficult to tell when or if a liquid culture is fully colonized. When using barley malt sugar, the liquid will discolor slightly from the color of the sugar. This makes it difficult to see the culture growing. If you're using light corn syrup, the solution should be completely clear, and you'll be able to clearly see a white culture growing in the solution.

In all instances when creating a liquid culture, I test it. After it has colonized, I usually draw up several syringes of the fluid and test them on a jar of grain or an agar plate to ensure that they grow properly and are not contaminated. At the 2-week mark, I put the liquid culture into the refrigerator. Two weeks is plenty of time for the culture to become fully colonized. Liquid cultures can easily last 6 months to a year if they are refrigerated.

Making Larger Batches

As you realize that you need more and more liquid culture solution, you can simply move up to a larger size canning jar — from half pints (250 mL) to pints (500 mL), and eventually from pints to quarts (1 L). Once you get to quarts, prepare them just as you would your grain jars, with a filter in the lid.

More experienced growers often step up to preparing their liquid cultures in 68 oz. (2000 mL) or larger conical laboratory flasks, also known as "Erlenmeyer flasks." These large containers need to have air circulated into the culture to help promote growth. This is achieved by adding a magnetic stir bar to the flask before sterilization, and colonizing the flask on a magnetic stir plate. This continually mixes oxygen into the culture while it's colonizing. Erlenmeyer flasks are inexpensive, but magnetic stir plates can be quite expensive. Look for them at online auction sites.

Once you've mastered the basic technique, it's not difficult to increase production of liquid culture by using an Erlenmeyer flask and a magnetic stir plate.

Inoculating with Agar

When creating liquid cultures with colonized agar petri dishes, it's best to chop the agar up into small pieces before adding it to your liquid culture medium. This distributes fragments of mycelium more thoroughly throughout the solution, and decreases colonization time. The most expensive piece of equipment I own is not my flow hood, as you might expect, but rather a blending container used for chopping up agar for use in a liquid culture. Most blending containers are made of plastic and would not withstand the high temperatures of a pressure cooker. There is a company called Eberbach that makes stainless steel blending containers that will withstand pressure cooking. The downside is that they cost more than $700.

There is a cheaper alternative that I wish I had known about when I started growing mushrooms. Several companies make blender blade bases that can be attached to quart canning jars with a standard ring band. These cost less than $20.

Making a Large Batch of Liquid Culture

Here's a method for making 2.5 pints (1250 mL) of liquid culture solution from one colonized agar petri dish.

Materials

- Colonized agar petri dish (see chapter 11)
- Stainless steel blender container or blade
- Blender base
- Pressure cooker
- Scalpel
- 2000-mL Erlenmeyer flask
- Light malt extract or light malt sugar
- Poly-fil or nonabsorbent cotton
- Aluminum foil
- Magnetic stir bar
- Magnetic stir plate
- Flow hood

1 **Prepare the malt extract.** Fill your flask with 750 mL of water and 50 grams (1.8 ounces) of light malt extract. Add your magnetic stir bar to the flask. Stuff the opening of your flask tightly with nonabsorbent cotton. Cover the top of the flask with aluminum foil.

1

2 **Sterilize your equipment.** Put 500 mL (17 ounces) of water into the blending container. Pressure cook the blending container, the flask, and a scalpel wrapped in foil for 45 minutes at 15 psi. Remember to heat the cooker slowly. I often pressure cook another canning jar of plain water in the cooker as well, in case any of the water in the flask is expelled during sterilization. Once the pressure cooker cools down to near 0 psi, move the cooker in front of your running flow hood. This will prevent any contaminants from being drawn into the cooker while it cools.

3 **Blend the colonized agar.** Once the blending container is completely cool, unwrap your scalpel and transfer your colonized agar plate into the steel container. Blend the agar and water mixture. I normally blend for 5 seconds, turn it off for 5 seconds to let the contents settle, blend for 5 seconds, wait again, and end with one more short blend.

3

4

5

4 **Combine agar and flask mixture.** Remove the foil and cotton from the top of the flask, and pour your blended agar dish into the flask.

5 **Wait for colonization.** Put the cotton back in the opening of the flask and recover with the foil. Place your flask onto your magnetic stir plate. Wait 1 to 2 weeks for the culture to colonize.

6 **Extract your culture into syringes.** I normally sterilize an empty quart jar and pour the contents of the flask into the open quart jar in front of the hood. This way, you can easily make the syringes you need at the time and refrigerate the rest for future use.

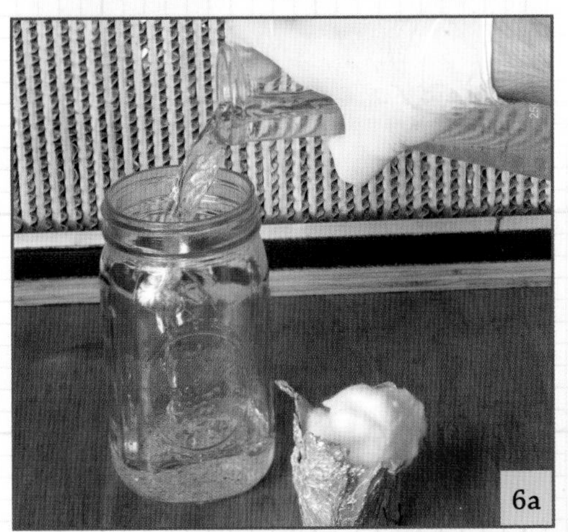
6a

6b

CHAPTER 9

WORKING WITH SAWDUST

The most common varieties of edible mushrooms include Shiitake, Lion's Mane, Maitake, and Reishi. All these edible species can be found growing from dead wood or tree stumps in the wild. This chapter describes how you can create the end substrate from which these mushrooms will fruit. You can simulate these large pieces of wood found in nature by filling up spawn bags with sawdust and mixing in the grain spawn you've learned to generate in previous chapters. There are two different methods that home growers are likely to use when working with sawdust: sawdust spawn and sawdust production blocks. Each method has a different purpose and procedure.

Sawdust Spawn

As you know from previous chapters, spawn is a substrate that you generate with the intention of transferring it to yet another substrate before you fruit it. Think of sawdust spawn as an intermediary step that will allow you to continue to expand your mycelial culture, while also allowing the growing culture to become accustomed to the sawdust long before it's asked to fruit. With any transfer of mycelium to a new substrate, there's a lag time during which the mycelia must learn the most efficient way to digest this new food.

Your culture will generally grow slowly after it's transferred, but its growth will speed up as it becomes more accustomed to the substrate. Speeding up the final outcome is one advantage of using sawdust spawn as the final step before creating sawdust production blocks. Another advantage of sawdust spawn is that it's much easier to generate than grain spawn and less likely to suffer contamination. Sawdust spawn can also be useful if you're creating mushroom logs outdoors (see chapter 2) or if you're generating your own plug spawn.

Types of Sawdust

The type of sawdust available to you depends largely on where you live. The most common edible mushrooms, like Shiitake, will grow on a variety of different hardwoods, but most outdoor growers look for oak logs to grow their mushrooms. For indoor growers, oak sawdust works very well. Either white oaks (including chestnut oak, white oak, chinkapin oak, etc.) or red oaks (pin oak, black oak, and red oak) perform well when working with hardwood-loving species, but other types of hardwoods will also work.

If I have a choice, I usually look for one of the "softer" hardwoods, such as tulip poplar. Less dense than the oaks, it will decompose more quickly and fruit your mushrooms faster. Most outdoor growers choose oaks for the opposite reason: it takes an oak log several more years to decompose than one of the softer hardwoods. Because the log lasts for a longer period of time, the outdoor grower receives more harvests. Indoors, we want the block to decompose more quickly so we can receive the harvests faster.

Aside from oaks and poplar, most of the other hardwoods will work fine as

Sawdust is an excellent substrate for mushroom cultivation. Be sure to use sawdust that is medium-textured — neither too fine nor too coarse.

well. These include, but are not limited to, ash, beech, birch, alder, blue beech (musclewood), elm, ironwood, sycamore, sweetgum, and willow. Some growers will also use sassafras, but I tend to shy away from the more aromatic woods.

Sawdust varies by particle size. Some sawdust is very fine, almost like a powder. It generally comes from mills cutting veneer or working with furniture. Sawdust can also have a very coarse texture, almost like small wood chips. Both of these extremes should be avoided. Sawdust for mushroom cultivation shouldn't be too fine, but it shouldn't be too coarse, either.

Sources of Sawdust

Finding a steady source of useable sawdust may be more difficult than you think. People used to think of sawdust as a worthless byproduct of the lumber industry, but that's no longer the case. Many modern sawmills sell all the sawdust they produce as bedding for horses, hogs, or cattle. Where I live, in southern Indiana, there are large swaths of public and private forests that are logged regularly. Sawmills in areas like these are a good first point of contact. If you're only looking for a small amount of sawdust, many mills will simply fill a few plastic totes free of charge. Procuring larger amounts may take a bit of negotiation.

Keep in mind that different sawmills have different procedures. Some cut certain species of wood on specific days, so they might have a pile of sawdust from a single tree species that you could acquire. Others intermix the tree species they are cutting, so the sawdust pile may contain remnants of many different tree species. For the small home grower, a mix of sawdust shouldn't really be a problem, as long as it's a mix of hardwoods. Larger mushroom-growing operations may prefer single-species sawdust to maintain quality control and predictable growing timeframes.

Sawdust from specialty furniture manufacturers is generally very fine, kind of like a powder, and isn't ideal for mushroom cultivation. The smallest grain size you should be looking for is about 1/32 inch (1 mm), the consistency of oat or wheat bran. Some sawmills may also distinguish sawdust from what they call hardwood "shavings." The individual pieces of these shavings tend to be a bit larger than particles of sawdust, but will still work fine for our purposes, especially if you can mix them in with smaller-sized sawdust particles.

If you don't live in an area that has a lumber industry, there are other options to explore. Farm and feed stores often sell bags of sawdust that range from 50 to 200 pounds (23 to 91 kg), primarily as animal bedding. Some manufacturers indicate the tree species included in the sawdust, but usually it's simply sold as "hardwood sawdust," and it's generally fine for our purposes. Many garden centers also sell sawdust in bags or bulk form.

Hardwood pellets — used in pellet stoves — can be used as a substrate material. When they're soaked in water, they break down into sawdust.

If you have trouble finding sawdust in an urban setting, the best alternative is wood pellets. Used in wood-burning stoves, wood pellets are usually available in 50-pound (23 kg) bags in big box and hardware stores. Wood pellets are just sawdust that has been heated and compressed; when you add water, they break down. Before you buy them, however, it's best to call the manufacturer or check the website to confirm that glues or other

No Walnut, Please

One tree species you should make sure is not included in your sawdust is walnut. Walnut trees produce a substance called juglone that is toxic to many plants, and can cause allergic reactions in animals and humans. Most gardeners and farms specifically request sawdust with no walnut, and mushroom growers should avoid it as well.

adhesives haven't been added during the manufacturing process. Most wood pellets contain only oak sawdust, so they should be great for growing mushrooms.

Finally, you can look for sawdust online, especially at sites like Craigslist, where sawdust is commonly offered for sale in units of 15 to 20 cubic yards (11.5 to 15 m^3). This means that the seller, usually a company that buys sawdust from a lumber mill, will arrive in a fully loaded dump truck, so keep that in mind when you're shopping. Most of these companies will also let you pick up smaller amounts at their place of business for a nominal fee. They may be able to get sawdust of a specific species of tree for you, so don't hesitate to call and ask.

Spawn Bags

A spawn bag is a sealed, side-gusseted polypropylene bag that has a filter patch near the top that allows for air exchange. A cheap and efficient way to sterilize large amounts of substrate, a spawn bag is one of the few items that is vital to the mushroom cultivation process, but is impossible to obtain locally. They are used not only for generating sawdust spawn, but also for grain spawn (chapter 6), plug spawn (chapter 2), and many other substrates. Some people pasteurize compost or manure in them as well (chapter 13). Many companies sell them online, and most of them originate from just a few manufacturers.

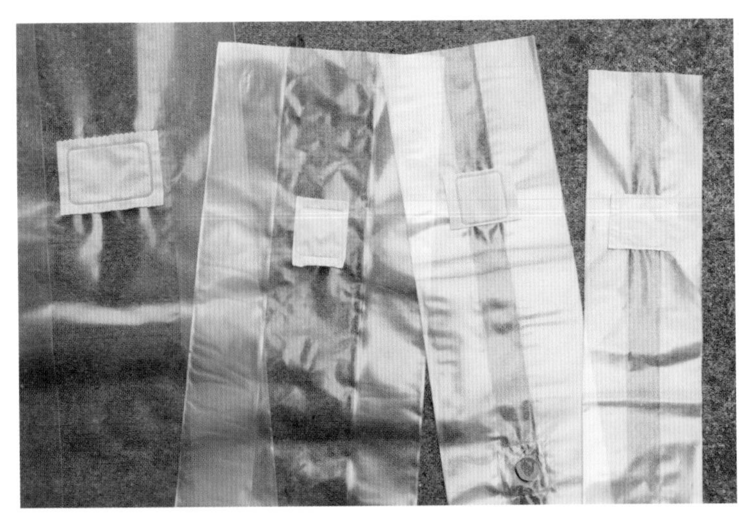

Spawn bags come in different sizes for different uses.

Spawn bags are generally sold in two sizes: medium and large. Both sizes are roughly 18 inches (46 cm) tall, while the medium bags are 4 inches (10 cm) wide and the large bags are 8 inches (20 cm) wide. Either size will work well. Most professional mushroom farms use the larger bags when preparing their substrates and in premade Shiitake mushroom kits. I sometimes use the medium-sized bags when preparing grain spawn.

The one variation you may encounter between spawn bags from different vendors is the type of filter molded into the bag. Different types of filters allow different-sized particles to pass through. The most common "efficiencies" are 0.2-micron, 0.5-micron, and

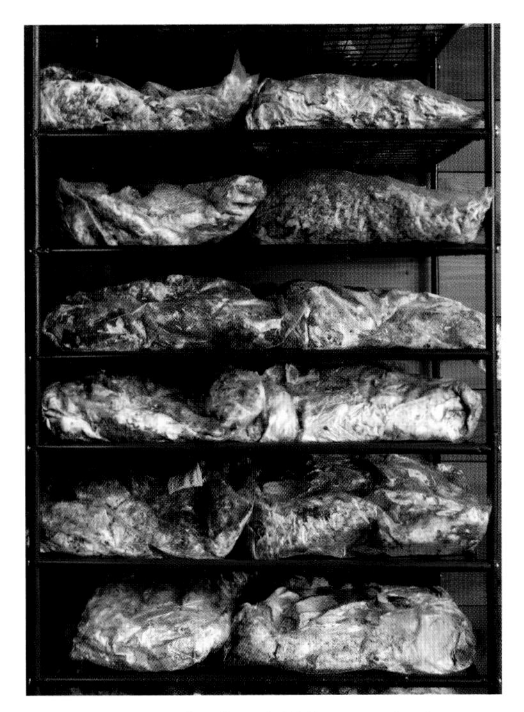

Sawdust spawn is often laid flat to colonize.

5.0-micron filter patches. When you're working with sawdust spawn, stay away from the 5.0-micron filter patch bags, as a filter of this efficiency has holes large enough to allow many of the most common contaminants into the substrate. The 0.2- and 0.5-micron filters are efficient enough to stop most potential airborne contaminants.

The advantage of the 5.0-micron filters is that you can seal them completely before sterilizing your substrate. If you completely seal the bags with the more efficient filters, the bags will probably burst during the sterilization process. If you consider using 5.0-micron filters, keep in mind that they allow particle sizes large enough to cause contamination in your cultures. They are better suited for use with substrates that require pasteurization rather than complete sterilization.

Spawn bags are also available in thicknesses ranging from 2.2 mil to 3.0 mil. I generally use the thinner bags and have always had success with them. The thicker bags cost a little more, and some people find them easier to work with. If you find that the thinner bags stretch or puncture during the sterilization process in a pressure cooker, you may want to try the thicker bags.

If your bags melt or become malformed, there are a several things you can try before switching bags. Spawn bags may burn through during sterilization if you heat the cooker up or cool it down too quickly. Sharp changes in temperature and pressure can cause the bags to become malformed. If the bags come into direct contact with a part of the cooker that's being heated, the high temperature can melt them. To solve this problem, raise the standoff plate a little higher from the bottom of the cooker. I often use old ring bands from jars to accomplish this. You can also try wrapping a small hand towel around the spawn bags in the cooker so they don't come in direct contact with the metal on the sides.

Sealing Your Spawn Bags

One final piece of equipment I recommend for working with spawn bags is an impulse sealer. Designed to create an airtight seal that keeps out potential contaminants, impulse sealers melt the two walls of the spawn bag together by exposing them to a thin strip of heated metal. After pressing down on an impulse sealer, it should only take 2 or 3 seconds to get the airtight seal you need. They come in a variety of sizes (8 inches, 12 inches, 16 inches, etc.) but the 16-inch model works best with large spawn bags. Most mushroom supply companies carry impulse sealers, but they can also be readily found on eBay for a great price.

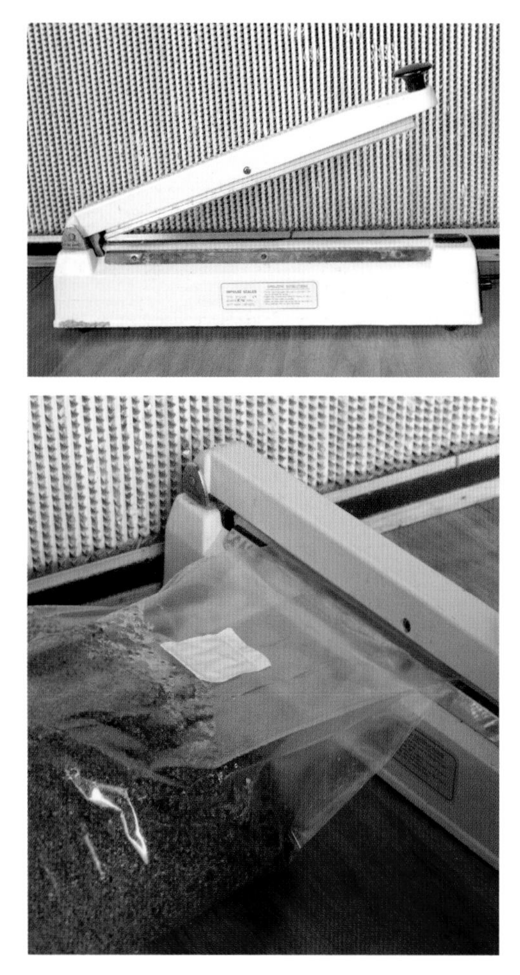

An impulse sealer is an important time-saver for the home cultivator.

Some people use vacuum sealers if they can't find an impulse sealer. If you get one, you'll be able to vacuum seal foods as well as spawn bags, but I tend to avoid dual food–mushroom products in my home, because moving them repeatedly in and out of the spawning area increases the chances for contamination.

One last-resort option is to use tape to seal your spawn bags. I did this for a month or so for small amounts of spawn with good results, but it's very time-consuming and not recommended if you want to obtain consistently reliable spawn. I simply folded 2-inch-wide (5 cm) wide masking tape over the top of the bag, and then put one additional strip on each side. If your bags are wet, the tape will not stick very firmly and your success rates will go down.

Bag clamps are another option. Like tape, these are probably not ideal for most circumstances, but can be used if necessary. Made by the same manufacturer as the spawn bags, bag clamps are simply two pieces of plastic that clamp to form a seal over the walls of the bag. They're not widely used and can be hard to find.

Now that you have the necessary background on the materials and equipment, here's how to prepare small amounts of sawdust spawn. Two 5-gallon (19 L) buckets of sawdust, with water, make about eight bags. There are many other ways to accomplish the same goal, but this is how I do it.

Bag clamps are specifically designed for mushroom production.

As a last resort, tape can be used for sealing spawn bags.

Finding Sawdust's Current Moisture Level

This is a very easy process that uses two weight measurements and an oven.

1. First, record the weight of a sample of your fresh sawdust. The exact amount doesn't matter; a handful will work well. This will be your wet weight.
2. Put your sample in the oven for an hour or two at the lowest temperature setting or until the sawdust is dry, then weigh the dry sample. This will be your dry weight.
3. Use the following equation to find the moisture content of your sawdust:

$$\text{Moisture Content} = \frac{\text{Wet Weight}}{(\text{Wet Weight} + \text{Dry Weight})}$$

Creating Sawdust Spawn

Materials

- Sawdust, one 5-gallon (19 L) bucket for every 4 spawn bags
- Water
- Measuring bowl or bucket
- Scale
- Spawn bags
- Clothespins
- Pressure cooker
- Glovebox or flow hood

Time Required

20–30 minutes for prep; 2–3 hours for sterilization

1. **Weigh the sawdust.** Weigh the bucket alone on the scale. Fill the bucket to the top with sawdust and weigh it again. Subtract the bucket weight from the total weight and you have the total weight of sawdust per bucket. My 5-gallon bucket contains about 9 lbs. (4 kg) of sawdust.

2. **Hydrate the sawdust.** Add water to get your sawdust to a moisture content of about 60 percent. To do this, you'll need to figure out how much water to add to the sawdust.

   Five pounds of hydrated sawdust at 60% moisture should have 3 lbs (1.4 kg) of water and 2 lbs (0.9 kg) of sawdust (5 × 0.6 = 3).

   Using this calculation, you can determine how much water you need to attain the desired moisture content for any given weight of sawdust. Ten pounds of sawdust needs 15 pounds (6.8 kg) of water. Keep in mind that this will only work for dry sawdust. If your sawdust is very fresh, you'll need to find out its current moisture level and subtract it from the 60 percent. That process is explained on page 159.

   After you know what weight of water to use, fill a bucket on a scale with that amount. The first time you do this, make a permanent mark on the interior of the bucket with the correct amount of water to use. That way, instead of using a scale every time, you can just fill the water up to that line in the bucket. Mix the water thoroughly into the sawdust.

3. **Fill the bag.** Fill a spawn bag with approximately 5 lbs. (2.3 kg) of moistened sawdust. Remove as much air as possible from the spawn bag and fold the top of the bag over two or three times. Put a couple of clothespins on the bag.

4. **Pressure cook the bag for 2–3 hours.** I normally use 2.5 hours for a large spawn bag containing 5 lbs. (2.3 kg) of sawdust. Once the pressure cooker is cool, transfer a colonized quart jar of grain spawn to the sawdust using a glovebox or a flow hood.

Inoculating Sawdust Bags

Inoculating sawdust spawn is a fairly simple process. You start with colonized quarts of grain (chapter 6) or bags of grain spawn (chapter 12) that are broken up. (See page 113 for tips on breaking up quarts of grain spawn.)

Have your spawn, sterilized sawdust block, and tools for sealing your bag in front of your flow hood or in your glovebox. Before opening the jar of grain spawn, wipe down the exterior with an alcohol swab or spray it with Lysol. Remove the clothespins from the bag and open it up, exposing the interior sawdust substrate. Open your jar. If you're using a medium spawn bag, pour half of the contents of the quart jar into the bag. If you're using a large spawn bag, use the entire quart. If you're using sawdust spawn, pour in similar amounts by volume. Once your spawn is in the sawdust bag, seal the bag.

After you seal the bag, give it a gentle squeeze with both hands. If the seal doesn't pop and the bag doesn't deflate at all, then the seal on the bag is good. If the bag pops, you know there was a weak spot in the seal, and you can simply seal it again. If the bag seals improperly, you may need to adjust the heat setting on your sealer.

If you're using a flow hood, you can position the open mouth of the bag to face the filter, which will force air into the bag as it seals. Allowing a pocket of air in the bag will make the next step, shaking the bag, much easier. You can also seal the bag facing sideways, as I do.

Once the bag has a good seal, set it off to the side and repeat the process to finish your bags. When all are finished, shake the bags thoroughly and place them in your incubation area. I generally lay sawdust spawn on its side and keep production blocks standing up for colonization during this period. Refer to the species pages (see pages 21–24) for optimum colonization temperatures.

Production Blocks

Although making sawdust spawn is a way to increase the size of your mushroom culture and to begin to acclimate your culture to its final substrate, sawdust production blocks are used to produce mushroom fruitbodies. Another name for these are "supplemented" sawdust blocks, referring to the addition of nutrients like oat bran or wheat bran, which will increase the yields of your final crop.

The process for creating production blocks is not very different from the process for creating sawdust spawn; the only difference is the addition of a nutritional supplement. The most common supplements, oat bran and wheat bran,

Notes and Tips for Production Blocks

Some growers suggest using wood chips as a part of your fruiting formula. These larger chips help increase oxygen availability within the block and increase your yields. In my experience, a good supply of fresh hardwood chips is more difficult to obtain than sawdust, but if this isn't the case in your area, you may want to take his suggestion. Add hardwood chips in the size range of 0.5 to 4 inches (1 to 10 cm) to your supplemented sawdust mix. Paul Stamets uses a mixture of 50 percent woodchips by weight and/or volume in his mix. So to 10 pounds (4.5 kg) of sawdust, add 5 pounds of woodchips (or one 5-gallon bucket of woodchips for every two 5-gallon buckets of sawdust). Adjust your water and bran ratios accordingly.

A garage with a smooth floor is the best place to work with sawdust at home. Most other processes can be accomplished with little mess in a kitchen or living room, but sawdust tends to get everywhere, and is impossible to clean. You'll generally be working with large amounts, and it will need to be thoroughly mixed and then loaded into bags, which requires a significant amount of space. On a smooth garage floor, you can lay down the sawdust, add water and/or bran, and mix it all using a large shovel. It can be easily cleaned up with a broom when it has dried.

If you don't have a garage floor, mix your sawdust in a large yard bin with a shovel. A large bin will allow you to make six or eight bags in one batch. A household cement mixer is another option; they come in many sizes and a basic model costs $300 to $400.

are available in 5-, 10-, 25-, or 50-pound (2.3, 4.5, 11, or 23 kg) bags at health food stores. The larger quantities may have to be ordered. If health food store prices are too high for these grains, a 50-pound bag of wheat bran is probably available from a feed store for less than $20.

If you add bran to the mixture, it should equal approximately 20 percent of the total mass of the mixture. For example, if you're using 10 pounds (4.5 kg) of sawdust, you'll need to use 2½ pounds (1.1 kg) of bran (2.5 / 12.5 = 20%).

You'll notice that gypsum is listed as an optional addition to your sawdust mixture. Renowned fungus advocate Paul Stamets cites two studies that show that gypsum stimulates mycelia growth and aids fruitbody formation

This Oyster production block is fully colonized and ready to fruit.

Shake production blocks well after inoculation.

"Popcorning" is a growth form on Shiitake production blocks that indicates the block is almost ready to fruit.

and development. Stamets suggests adding 2 to 3 percent gypsum by weight, or about 1 percent by volume. Both of these numbers exclude the weight and volume of water.

The Next Steps

The procedures for cultivating every hardwood-loving mushroom species I know are identical up to this point. But the temperatures at which they are incubated, the method of fruiting, and the fruiting temperatures vary by species. Some species do well in warm weather; some thrive in colder weather. Fortunately for home growers, most will do exceedingly well at room temperature. With Shiitake, you'll remove the entire spawn bag for fruiting; for Lion's Mane and Oyster mushrooms, you'll cut slits into the sides of the colonized spawn bags; for Maitake, you'll simply cut off the top of the bag and the fruits will form from the top of the block. To finalize your grow, you'll need to refer to the individual species pages (see pages 21–24).

Shiitake mycelium naturally browns as it ages. This block is ready to enter the fruiting chamber.

CHAPTER 10

MIDSIZE FRUITING CHAMBERS

The size and design of your fruiting chamber depends on the type of substrate you're trying to fruit and the space you have available. In chapter 4, we discussed small-scale fruiting chambers made from plastic storage bins, designed to house small amounts of mushrooms. This chapter explores larger fruiting chambers, requiring space equivalent to a small bedroom closet.

It's possible to produce more mushrooms simply by using larger storage bins, and some people prefer to stick with setups they know work well; however, this tends to become less than ideal as you add containers, because of the large amounts of perlite required and the frequent need to move stacked bins around. Perlite is very messy to work with, so I was especially anxious to give it up as quickly as possible. My main suggestion for a midsize fruiting chamber is a variation on a four-tier greenhouse design.

Shiitake blocks enter the fruiting chamber — in this case, a four-tier greenhouse.

The Four-Tier Greenhouse

These greenhouses are probably the most versatile fruiting chambers available for the home cultivator. Most garden centers at big box stores carry them, and they are also available online. Try to locate one locally if you can, as shipping can be expensive. The best time of year to buy them is in the late fall when they are often deeply discounted.

These greenhouses can be easily modified to fit many different growing styles and locations. I have used them in basements, closets, and spare bedrooms.

Setup

The greenhouse is essentially a four-shelf wire rack with a plastic outer covering. The bottom is an open wire rack, so the first thing to do is to lay some plastic down on the area you've selected for the greenhouse. You can put it in an area with carpet, as long as there's plastic covering the floor. A large black plastic garbage bag should be just the right size, but you may want to lay plastic over the entire grow area.

There are two different greenhouse models available. The most common model has plastic joints and is assembled by pressing metal side rods into the plastic frame. A second model is constructed completely of metal; metal shelves screw directly into metal side rods. All-metal greenhouses are superior: they typically last much longer. Plastic

All-metal greenhouse stands (left) are much sturdier than the plastic ones (right) and are worth the higher cost.

joints tend to break, especially when disassembling the greenhouses for storage or transport. The all-metal models can withstand being disassembled and moved many times, so it's probably worth a little extra money.

After your greenhouse is assembled, the only major concern to address is the grow environment. As with all fruiting situations, the main variables are humidity, temperature, light, and fresh air.

Humidity

One of the primary reasons a grower might choose to upgrade to a midsize fruiting chamber is because it allows for easier automation of important processes. Humidity is a major concern. Smaller fruiting chambers rely on a combination of direct spraying and evaporative humidity from perlite to maintain appropriate humidity levels. Midsize chambers rely instead on a direct injection of humidity from a humidifier. It's fairly easy to add a humidifier to a greenhouse and to adjust it so you never have to use a spray bottle again.

Two types of humidifiers are generally available: cool mist and ultrasonic. Both can work in greenhouses, but I believe the ultrasonic humidifiers are far superior. Because they attain the proper humidity level faster, they don't have to run as long. And you can actually see the humidity flowing from the unit, so you have a visual gauge of how much humidity is in your greenhouse at any one time.

Ultrasonic humidifiers should be operated using only bottled spring water. They aren't really designed for long-term hard use, and if you use normal tap water, minerals and other types of buildup will start to accrue on the atomizer, eventually causing the unit to fail. Using only bottled water will double or triple the life expectancy of your humidifier.

Humidifier Setup

The most common way to operate a humidifier with a greenhouse is to place it next to the greenhouse and to pump the humidity in through some kind of tubing. Placing it on the outside will force fresh air as well as humidity into the greenhouse, helping with air exchange. If you place the humidifier inside the greenhouse, you'll

With midsize fruiting chambers, it's easier to maintain appropriate humidity levels by attaching a humidifier, rather than relying on spray bottles.

just be recycling the air in the greenhouse, which will allow CO_2 to build up.

I always set the humidifier on the floor next to the greenhouse. This may not be ideal, as most potential contaminants are closest to the ground, but I have run setups like this for years with no significant contamination issues.

A single humidifier should be sufficient to run two or more greenhouses; I've humidified four greenhouses from one humidifier with great success. I find PVC tubing the easiest material to use for directing the humidity from the humidifier into the greenhouse. Many other materials will also work, however, so feel free to work with whatever you have available.

Measuring Humidity

A simple digital humidity gauge, which costs less than $20, should allow you to monitor humidity level within ±5 percent relative humidity. This level of accuracy should be sufficient for your needs. These inexpensive units are usually not accurate above 90 percent relative humidity, and may simply read "High" if the relative humidity rises above the proper range.

Here's a little trick that will allow you to check the relative humidity without

Using a combination thermometer/hygrometer is an easy and inexpensive way to track the temperature and humidity of the fruiting chamber.

a gauge. Spray the walls of your fruiting chamber with a mister, and come back in 3 to 4 hours. If the wall is still moist where you sprayed, the humidity is near the right range. If the wall has dried out, you need to increase the humidity level a little bit. If the walls are dripping wet with humidity, your levels are too high and should be decreased.

Many mushrooms are more productive when they're provided with a constant and even light source. Shop lights work well in this application.

Air Exchange

When using the four-tier design, a good portion of your air exchange comes from the air forced in through the humidifier. This may not be sufficient, however. It's still best to open the greenhouse once a day or every other day to air it out. Some people add small computer fans to the sides of the greenhouse and put them on a timer, but I've never found this necessary.

Finally, I suggest that you do not seal the bottom of the greenhouse to the floor. Leaving the bottom of the greenhouse loose on the plastic underneath allows a little more air to circulate throughout the day.

Light

When you increase the size of your fruiting chamber, you also need to increase your light sources. Although most types of mushrooms don't require much light to grow, many grow better if they have a constant and even source. These four-tier greenhouses work well with one shop light per two greenhouses. The 4-foot (1.2 m) shop lights work extremely well when you install them in the vertical position shining onto the greenhouses. This allows every level to get an even amount of light. I suggest using the same lighting timeframe for all species: 12 hours on and 12 hours off. High-intensity lights are never necessary for mushroom cultivation.

Temperature

The final main greenhouse consideration is temperature, and it may also be the easiest to maintain with this setup. It's as simple as maintaining the proper temperature of the room. If you need some heat, a small space heater is often sufficient. If you need it cooler, consider a small air conditioner in the room, or drop the temperature of your home or apartment by several degrees.

Growing

Once you have the basic setup done, these greenhouses work sort of like an oven. You put your finished substrate into the unit, set the time (usually about 2 weeks), and don't mess with it again until it's done. All you need to do is monitor your four fruiting variables: humidity, temperature, light, and fresh air.

Another advantage of these greenhouses is that they are easily scalable. If you start out with two in a closet and realize that you need more space, it doesn't take much effort to set up an entire bedroom with 10 or 12 greenhouses using the same design.

Other Options

Portable garden rooms also work as self-contained greenhouses. Primarily used by hydroponic growers, they also work well for mushroom cultivation. Attachment areas are already built in for lighting and ventilation, the inside walls are lined with reflective Mylar, and they are designed to be waterproof.

Portable garden rooms are manufactured by several different companies and come in a variety of sizes; however, you'll need to find racks for the interior because they are not provided, and you'll want to maximize your growing area.

PART 3

Advanced Methods

CHAPTER 11

AGAR CULTURES

Working with agar is one of the most frequently discussed mushroom cultivation methods, but most home growers don't tend to try it until after they've tried many of the other methods. Before the public had access to the Internet, it was very difficult to find sources of inoculum. Now, with liquid cultures and spore syringes so readily and cheaply available, many beginners put off more advanced culturing techniques until they know they'll be sticking with the hobby for a while. But if you've grown mushrooms out completely several times and want to start maintaining your own culture collection, learning the lab methods associated with agar is a must.

What Is Agar?

Agar is a gelatinous material derived from seaweed. It's often sold in a powder form, but when mixed with water and allowed to solidify, it turns to a consistency similar to common gelatin. It has no nutritional value by itself, so it must be supplemented with a food source if mushrooms are to grow on it. Some of the most common nutritional additives used to formulate the final agar medium are malt extract and potatoes.

After you prepare this nutritious medium, you must sterilize the mixture in a pressure cooker and pour a thin layer into a container such as a petri dish or small canning jar. When the agar is hot, it flows like a liquid; as it begins to cool, it solidifies into a gel. Your mushroom cultures will grow out on the surface of this gel.

Learning to work with agar efficiently requires a significant amount of time and patience and a certain amount of trial and error. For the best chance of success, I recommend using a flow hood rather than a glovebox. A glovebox restricts your freedom of movement, slowing you down, and generally has much higher contamination rates when used with agar.

Why Use Agar?

Learning to work with agar is important for a number of reasons. The first is that you can use agar to isolate the most vigorous strains from a given species. If you inoculate a petri dish with spores, there may be literally hundreds of different viable strains growing on your dish at once, and not all of them will have the same growth characteristics. As they begin to grow out, the most vigorous and fastest-growing strains tend to dominate. Working with agar petri dishes allows you to cut away the healthiest sections of growth and transfer them to a new dish, a process called *strain isolation*. After several transfers, you're left with a single vigorous strain of a species that can be tested for other growth factors, such as fruiting time and fruitbody size, allowing you to find the optimal strains for your growing method.

Agar also allows for isolation away from contamination. Even the best mushroom spawn labs experience contamination from other molds or bacteria from time to time, but strains that show signs of contamination can be saved if you're using agar petri dishes. You can simply cut away the healthy growth and transfer it to a new dish. It may take several transfers before you achieve contamination-free dishes, but your contaminated cultures need not be thrown away and lost entirely.

A final benefit of working with agar is that you can build a collection of mushroom cultures and maintain it for a long time. Liquid culture syringes remain viable for 6 months to a year when kept in a refrigerator. After that, they need to be grown out and checked for viability. Agar cultures made in test tubes remain viable for many years if stored properly, potentially even decades. This allows you to build a culture collection of many different mushroom types, and to have each culture ready to grow again instantly if you decide to use it again.

Now that you know the benefits of using agar, you're ready to find out how to use it, including how to prepare media, inoculation, strain selection, cloning, and saving your strains for future use.

Agar allows you to develop and maintain a collection of cultures that can remain viable for years. They're simply produced with water, agar, and malt sugar.

Preparing Media

Within the world of microbiology, there are literally thousands of different types of agar media. Most are specialized mixtures for a specific purpose, such as optimizing or inhibiting the growth of bacteria or fungi. Within the world of mushroom cultivation, home growers tend to focus on two main agar formulations: malt extract agar (MEA) and potato dextrose agar (PDA). These two formulations will be the focus of this chapter.

Agar Containers

Whether you're purchasing or creating your agar media, one of your main considerations is the container that will hold your cultures. Here are a few things to keep in mind. First, the base of your container should not be much larger than 3.5 inches (9 cm) in diameter; a larger container is simply a waste of agar. Second, you'll have to choose between containers that can be sterilized and those that come in sterile packaging. Finally, your container needs a lid. If too much air is allowed to circulate into the dish, you increase the potential for the introduction of contaminants, and you risk drying out the agar and the culture you're trying to propagate. Following are a few of the many types of containers that can work.

Petri Dishes

Petri dishes, well-known in microbiology, are the containers most often used to hold nutrient-rich agar for culturing fungi and bacteria. Petri dishes can be made of glass or plastic.

Glass dishes are made of borosilicate glass, which means they can withstand repeated sterilization and be reused again and again. Most home cultivators use the 100 × 15 mm size. Easy to clean with soap and water, these dishes must also be steam sterilized without agar in them between uses. You can use sterilizer boxes or autoclavable racks to help with sterilization, but instead of purchasing this separate equipment, I stack six or seven empty dishes in a column and wrap them in aluminum foil before pressure cooking for 20 to 30 minutes.

Glass dishes take time to clean and sterilize between uses, but they save you a significant amount of money over time if you make dishes frequently; however, if you're like most home growers, you'll work with agar infrequently, so expensive glass dishes may not be your best solution. I used plastic disposable dishes for many years with great success. These polystyrene dishes measure 100 × 15 mm, just like the glass ones, and come sealed and sterile in a plastic bag of 20 to 25 dishes. Plastic containers melt in temperatures commonly attained while pressure cooking, so they cannot be reused, but in all other respects, they're just like glass.

Glass petri dishes (bottom) are more expensive than plastic ones (top), but they can be sterilized and reused for years. Plastic petri dishes cannot be reused.

There may be times when you open a bag of sterilized plastic petri dishes but only use a portion of the dishes. If this is the case, put the unused dishes back in the original plastic bag and tape it closed. The dishes will remain sterile enough to use in the future.

Canning Jars

Wide-mouth, half-pint canning jars (the same ones used for the PF Tek process discussed in chapter 4) also make good agar containers, and can often be found locally on short notice. Half-pint wide-mouth jars are shallower than most other sizes, which makes it easier to get a scalpel down to the bottom to cut out segments for transfer.

If you use canning jars when working with agar, don't poke holes into the lid as you do for the PF Tek process. Just turn the lids upside down on the jars so the seal faces upward away from the glass rim. This will allow for a small amount of air exchange but not enough to cause any significant contamination issues.

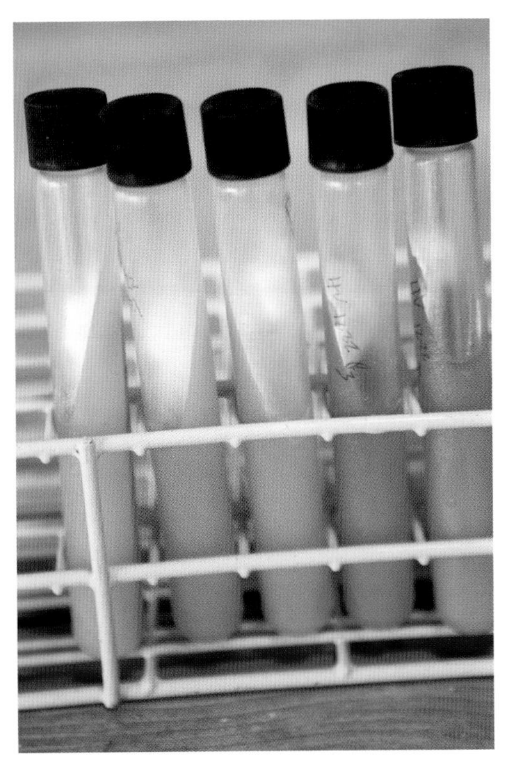

Agar slants are primarily used to store cultures long term.

Test Tubes

A test tube serves several functions when working with agar. Its primary use is as a vessel to store cultures for long periods. The creation of *agar slants*, a process by which the tubes are filled with agar and allowed to cool at an angle, increases the surface area of the agar within a very small space, and allows more room for the culture to grow out than if it were allowed to solidify standing straight up.

Very few cultivators use test tubes as their primary container for common agar work. The mouths of most test tubes are very small, and when you're working to clean up a contaminated culture, it's difficult to cut away the desired section with any amount of precision. There are many other container options, including small plastic drink cups or baby food jars.

Purchasing Media

The easiest way to ensure you have the proper mixture of ingredients in your agar is to purchase a premade mixture from a store. There are numerous stores online that offer premixed agar — a simple search for "malt extract agar" should yield numerous reputable companies. These companies often sell premixed agar in units of 250 or 500 grams.

To mix premade agar you'll mix slightly more than 3 tablespoons (20 to 25 g) of the powder with 500 mL (1 pt.) of water. This mixture can be placed in a quart canning jar or a large Erlenmeyer flask and pressure cooked at 15 psi for 30 minutes. Don't let the pressure cooker cool down completely before you pour your plates, as agar quickly solidifies into a gel. Instead, place the pressure cooker in front of the flow hood while it's still warm, and pour enough agar into each plate to cover the entire bottom of the container about ⅛ inch (3 mm) thick. You don't need a thick layer of agar in the dish, as your culture grows only on the surface. A layer that's too thick wastes agar; a layer that's too thin may cause the plate to dry out and your culture will be lost. For more information on pouring plates, see page 182.

After the dishes are poured, they should be left alone for several hours to cool and solidify. Mixing 500 mL (1 pt.) of this solution should allow you to fill 20 petri dishes.

Creating Your Own Media

While finding premade MEA at just any local store isn't easy, finding the components to make your own isn't difficult at all. Mixing up your own agar is a very simple process, involving very little additional time. All the formulas presented here are known to work well with most edible mushrooms. The one you choose depends primarily on your preference and the availability of materials.

The most expensive part of the mix is the agar itself. Many health food stores sell powdered agar in bulk, but it can be expensive. Grocery stores and Asian markets are also possible sources.

The next thing to look for is your nutrient source, as agar alone doesn't contain any of the nutrients required for your culture to grow.

Malt Extract

Malt extract in most of its forms, including light malt extract, dry malt extract, barley malt extract, and barley malt sugar, will work for our purposes. Malt extract is available in liquid form, but you need the dry version, which should have the texture of fine household sugar. Look for malt extract in brewing supply or health food stores.

Potato Dextrose Agar

This common formula is used for growing many types of fungi. Potatoes can be purchased in several different forms, and most will work for creating agar. The two methods discussed here involve whole potatoes and instant potato flakes.

Note on Storage

Be sure to store your extra malt extract in an airtight container. Malt extract readily absorbs water; if you leave it or your agar-malt mixture out in the open air, it will solidify quickly into a block because of the ambient humidity in your home. If your malt extract or agar mixture does harden, you can still break it up and use it. Just be sure that the particle size is as small as possible before mixing it with water for sterilization.

Malt Extract Agar

Materials

- 500 mL (2 c.) water
- 10 g (1⅓ tbs.) agar
- 10 g (1⅓ tbs.) malt extract
- 1 g (½ tsp.) yeast (optional)

I usually mix up a large batch of the dry components at one time, and then add 20 g (just over 3 tbs.) of the final mixture into 500 mL (2 c.) of water. So if I pour 200 g (7 oz.) of agar into a bowl, I add 200 g (7 oz.) of malt extract into the bowl as well and then save everything I'm not using at that time. That way, I always have my own premade mix on hand.

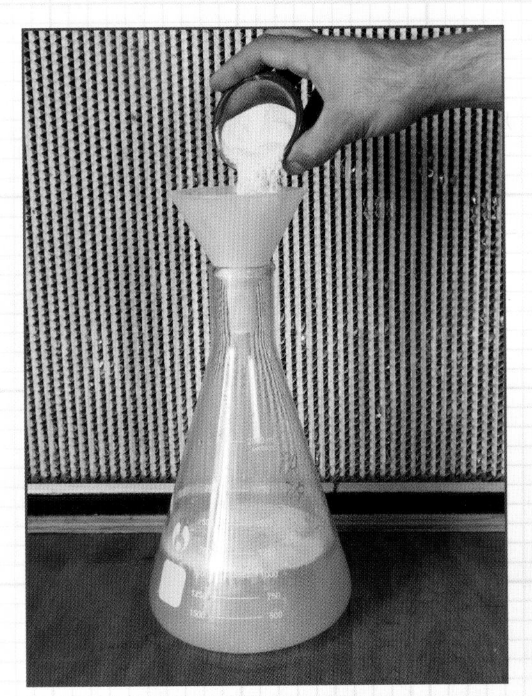

Creating Potato Dextrose Agar

Materials

- ½ to ¾ pound (227–340 g) unpeeled potatoes
- 1 L (4 c.) water, plus additional
- Pot for boiling water
- Cheesecloth or other fine strainer
- 20 g (just over 3 tbs.) agar
- 20 g (just over 3 tbs.) dextrose or dextrose substitute (light corn syrup)
- Pressure cooker
- Flask

1 **Start by creating potato water.** Slice the potatoes to allow more surface area to be in contact with the water; this results in more nutrients in the liquid. Place water and potatoes into a pot and boil them for 30–60 minutes.

1

2a

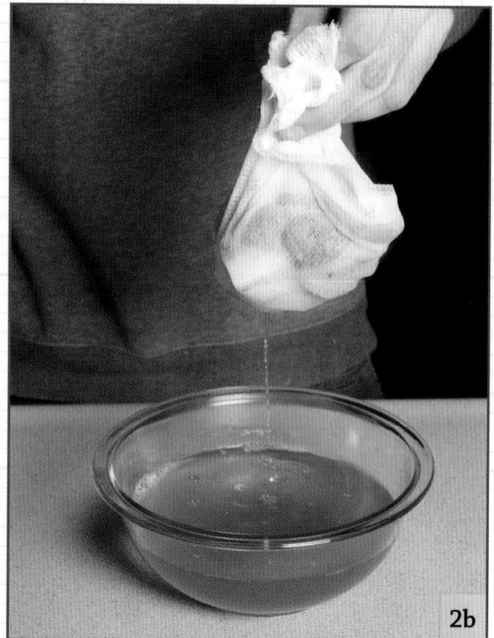

2b

2 **Strain the potatoes.** Using cheesecloth or another type of fine mesh strainer, strain the chunks of potato out of the water, but *save the water.* You can throw away the chunks of potato or save them for another use.

3 **Add agar and dextrose.** Add more water to bring the volume back up to 1 L (4 c.). Add agar and dextrose to the water. If you're using a light corn syrup substitute, you need about 4 tsp. Pressure cook the mixture for 30 minutes at 15 psi.

3

Potato Flake Agar

Potato flakes also work well to create your base of potato water, and this method is even easier than boiling potatoes. You can mix up a large batch of potato flakes and agar, just as you can mix batches of malt extract and agar, and store any extra for later use. Mix all the ingredients together and pressure cook as you would for potato dextrose.

- 500 mL (2 c.) water
- 10 g (1⅓ tbsp.) agar
- 5 g (1½ tbsp.) potato flakes
- 10 mL (2 tsp.) honey or light corn syrup
- 1 g (½ tsp.) nutritional yeast (optional)

Final Considerations before Cooking Agar

As you prepare to pressure cook your agar medium, think about the tools you will need to use; some of them may need to be sterilized before you use them. If so, pressure cook these tools along with the agar. The most common tools you'll need are petri dishes, a scalpel, and an inoculation loop.

Sterilize petri dishes only if you're using the reusable borosilicate glass dishes. You can buy special containers to stack petri dishes in for sterilization, but they're expensive and unnecessary. One efficient way to sterilize petri dishes is to wash them first if they've been used, arrange them in stacks of five, and then wrap them in aluminum foil. The foil wrap serves as another layer of protection against contamination after the dishes have been removed from the cooker and are waiting to be used.

Liquid agar can be sterilized in an Erlenmeyer flask (stuffed with Poly-fil and covered with foil), along with foil-wrapped tweezers and blender parts.

This foil wrapping method also works well for other tools; wrap them in foil before you put them into the pot. Scalpels are useful for isolating individual strains from the surface of your agar. You might need an inoculation loop to inoculate your dishes with spores. If you plan to inoculate your dish with living tissue from a fresh mushroom fruitbody, a process called *cloning*, tweezers are useful. A jar of sterile water and a shot glass may also be helpful with some cloning methods. Depending on the method you're using, an empty syringe may also be necessary. No matter what tools you're using, it's best to wrap them in aluminum foil and sterilize them with the agar media.

I generally sterilize my agar media in a 2000-mL Erlenmeyer flask. Its narrow mouth makes it easy to pour liquid agar into dishes without much spillage. To keep the media from boiling out during sterilization, I roll up nonabsorbent cotton or Poly-fil and stuff it into the opening. After the stuffing is tightly in the opening, I wrap a square of aluminum foil over the top of the container, using a method similar to the one I use with canning jars.

When sterilizing your agar in a flask, it's best to keep your cooker at a heat point that doesn't allow the jiggler to move. If the jiggler is rocking for the entire cooking time, the agar will be bubbling up in the container, and some may be ejected in spite of the stuffing. Keeping the cooker just below the point where it's jiggling will maintain a constant pressure in the vessel, and will not allow the contents of your media container to boil out.

Pouring Your Plates

After you have cooked your agar, the next step is to pour your plates. Agar straight from the pressure cooker will be very hot and in liquid form. Leave your agar sealed in the cooker until the pressure is down and you're almost ready to pour your plates. As the agar mixture cools in its container, it tends to suck in a small amount of air. Leaving the cooker sealed until you're ready to pour lessens the chances that contaminants will be sucked into the agar to infect your subsequently poured dishes.

Agar can also be sterilized in mason jars (along with stacks of glass petri dishes wrapped in foil).

The ideal time to pull out your agar mixture is while the pressure cooker is still very warm, but not hot to the touch. If you let the mixture sit too long in the pressure cooker, the agar will begin to solidify and you won't be able to pour it. For an All American 921 pressure cooker, this means you should pull the agar out of the cooker about an hour after you shut off your stove.

While the cooker cools, prepare your working surface. For me, this usually means firing up the flow hood for an hour before I plan on working, and wiping down my work surface with alcohol. I always wear rubber gloves when working with agar, even though it may not be entirely necessary, to eliminate as many opportunities for contamination as possible. Your hands are a significant potential source of bacteria and fungi. Wearing rubber gloves also allows you to easily rub alcohol over your hands whenever you touch an object that's not in your zone of sterile air (e.g., the pressure cooker lid).

The media container is usually too hot to handle when it's first removed from the pressure cooker, so let it sit in front of the flow hood for 5 to 10 minutes. Some people wrap a towel around the neck of the flask or jar, but remember that a towel is an unsterile object you're introducing into your workspace. You might be able to use a paper towel, as they can generally be thought of as fairly sterile, but always tear off and discard the outer piece of paper towel on the roll, as it has had extended exposure to unsterile air.

When your agar is cool enough to pour, either open up your sterile bag of plastic petri dishes or remove the foil from your reusable glass dishes. Put the dishes in your work area in stacks of three to five dishes, one on top of the other. Be aware that the sterilization process leaves residual moisture between the dishes that can make stacks of five slippery, and the top dishes can slide off.

Begin by taking all of the top dishes and the lid of the bottom dish in your hand. The only thing that should be left on the table is the bottom section of the bottom dish. This will be the first dish that you pour, and you'll work your way up the stack. When pouring your second plate, hold the top dishes and the lid of the second plate up the stack. Pour this second dish while it's sitting on top of the first dish, and repeat this process the rest of the way up the stack.

Pour just enough agar to cover the entire bottom of the dish with a thin layer about ⅛ inch (3 mm) thick. Your mushroom culture will be growing only on the surface; it won't grow down into the agar, so there's no need to pour a thick layer into your plates. Don't worry if some of your initial plates aren't so great; you may have to pour a dozen or two plates before you can pour agar quickly and efficiently.

If you don't quite get 20 petri dishes of the 100 mm size using 500 mL (1 pt.) of your agar medium, try pouring your plates a little thinner the next time. Once all your dishes are poured, leave them in front of your hood or in your glovebox for an hour or two until they completely cool down and the agar solidifies. It's common for condensation to form on the top of the petri dish lids while they cool down. Once the medium completely solidifies, you can move on to the next step.

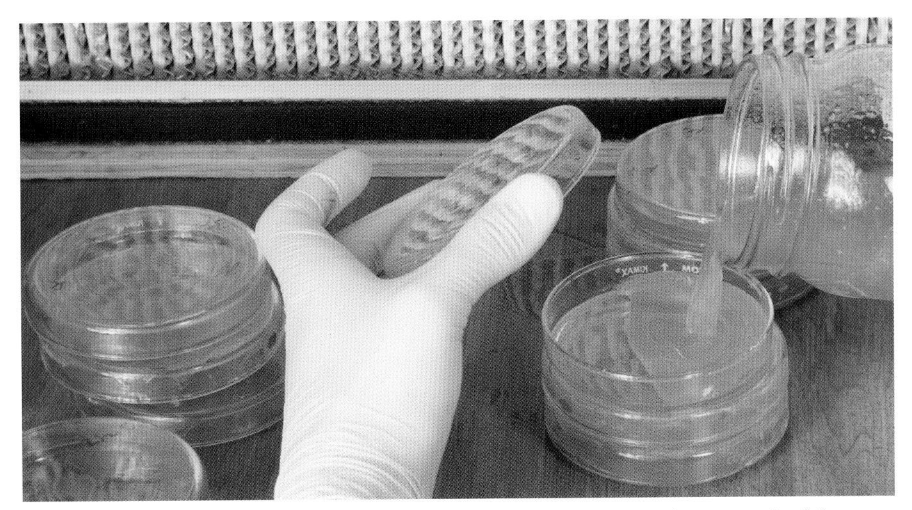

Once both the agar and the petri dishes have been sterilized, pour just enough of the agar to cover the bottom of the dish with about an ⅛-inch (0.6 cm) layer.

Inoculation

There are three main ways to inoculate an agar petri dish when starting a new culture. You can inoculate with spores, with living tissue (cloning), or with an already established culture in the form of an agar dish, an agar slant, or a liquid culture syringe. Each of these methods will be discussed in this section.

Starting with Spores

Contrary to popular belief, spores are not the primary method used for propagation of most mushroom species. More often than not, growers start with living mycelial cultures. But there are several important reasons why a grower may want or need to start with spores. The first is that spores remain viable for very long periods of time — more than 20 years in some cases. Storing spores as spore prints for these long periods requires far less maintenance or space than a culture collection. A second important reason to start a new culture from spores is to get some new genetics. Over time, a living mycelial culture can begin a process known as *senescence*, in which the ability of the culture to produce healthy, vigorously growing mycelium and fruitbodies begins to decline. With time and large numbers of cell divisions, the genetic viability of the culture decreases. Restarting a culture from spores will introduce new genetics to the culture, and using a process known as *strain isolation* (see page 187), a new and vigorous strain of a given species can be attained.

Once growers have a culture established from spores, they're not likely to return to the spores again for some time, because they generally continue propagating and then possibly cloning the initially isolated strain of the species. There are two common ways to start a mycelial culture using mushroom spores: spore syringes and spore prints.

Starting from a Spore Syringe

Creating your own spore syringe from a spore print was covered in chapter 4, so please refer to that chapter if you need to create your own. Spore syringes can also be purchased online.

If you're starting from a spore syringe, the process of beginning a culture is very simple. Once your agar is solidified on your petri dish, prepare to work in front of a flow hood or in a glovebox. Remove the cap that covers the needle of your spore syringe and, using a lighter, flame the needle of the syringe to sterilize it. Move your flame from the plunger end of the needle up towards the needle's tip. If you start at the needle's tip, the heat may force some of the fluid in the needle back up into the plunger. If any contaminants made their way to the open end, it will force them back up into the plunger. Starting the flame near the plunger and working your way out will force all the liquid currently in the syringe needle out of the opening near the tip, so if there are any contaminants, they'll be ejected from the needle.

After you flame the syringe, you'll need to eject a little bit of spore fluid into

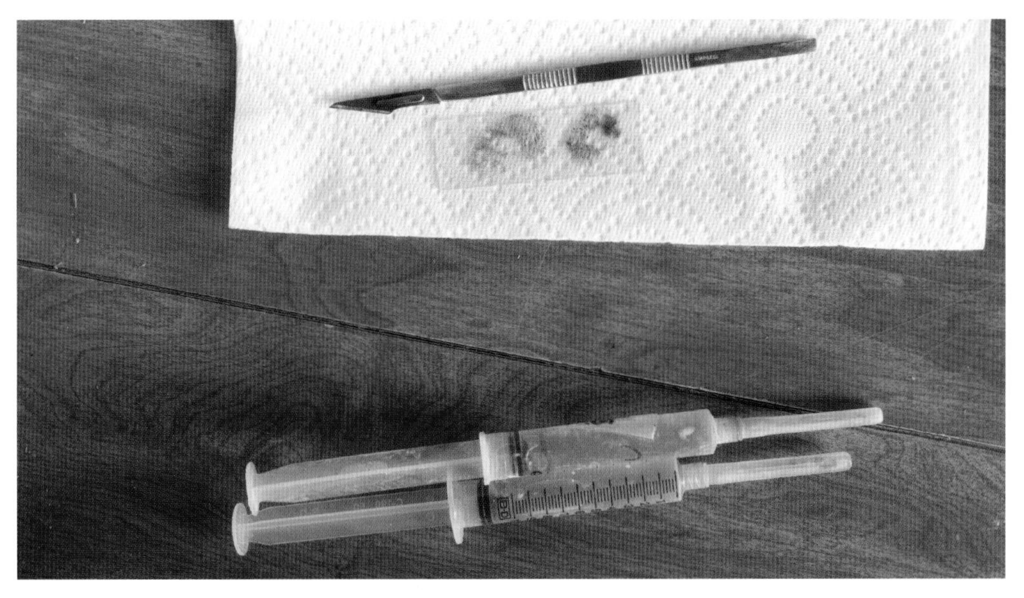

Agar can be inoculated with spores either from a spore print or a spore syringe.

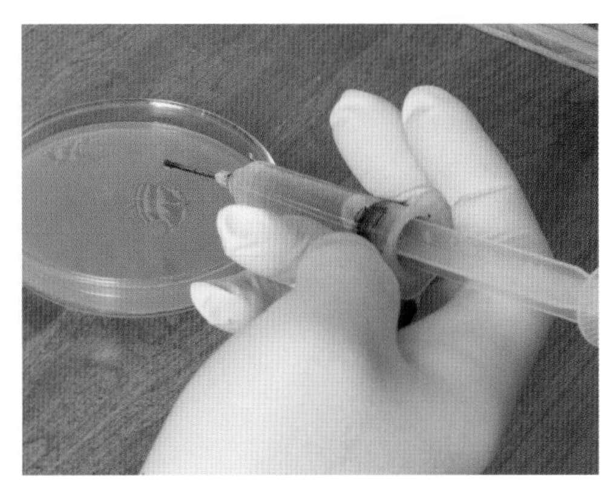

Only a few drops from a spore syringe will be enough to start a culture on agar.

a paper towel or an empty jar. This cools down the needle, so that fresh and viable spores can pass through the needle without being killed by the heat, and it leaves the needle filled with the fresh and viable spores you will use to start your culture. Another option is to cool the needle with a fresh alcohol swab. After the needle is cooled, simply eject a couple of drops of the spore fluid onto the center of your petri dish. It doesn't take much spore fluid to begin a culture — a drop or two per petri dish will suffice. Each drop of spore fluid contains a significant number of spores.

Starting from a Spore Print

Starting an agar culture from a spore print is also a fairly simple process. The only items you'll need are a spore print and an inoculation loop or scalpel. An inoculation loop is beneficial because it allows the spores to be streaked across the surface of the agar without cutting into the surface of the agar. An inoculation loop, available at lab and mushroom supply stores online, is a simple, inexpensive instrument: Basically, it's a wire with a loop on the end of a handle. You can also use a scalpel to start a culture from a print, but this method is not as precise and easy as using an inoculation loop.

To start, wrap your inoculation loop in foil and sterilize it when you pressure cook your agar mix. Set the wrapped loop beside your freshly poured dishes until they're cool. To inoculate the first dish, unwrap your loop and gently scrape the loop along your spore print. You don't need to have visible clumps of spores on the end of the loop, as that represents thousands of spores. I run the loop back and forth a few times over a small portion of the print rather than across the entire print. That way, if a portion of the print contains contaminants, you're less likely to pick them up for each dish. After you scrape the print, streak the loop along the agar surface of your petri dish, making an "S" shape across the dish. This spreads the spores out across a large portion of the surface and gives you lots of different sections of growth to choose from once your mycelium begins to grow.

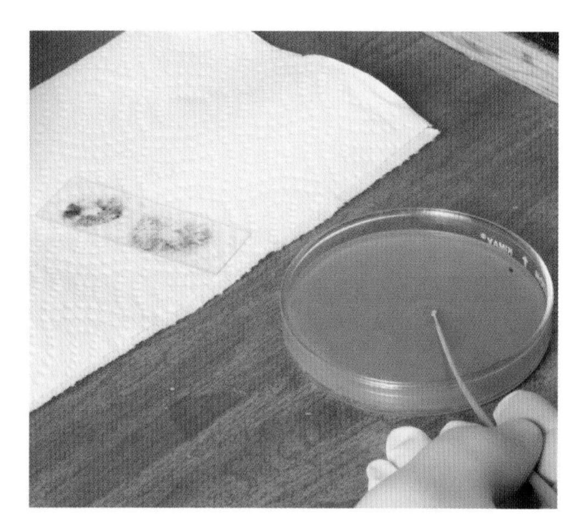

Transfer spores from a slide or piece of foil to the petri dish; in this photo I'm using an inoculation loop. Spread spores across the middle of the dish in an "S" pattern.

This plate of agar has been colonized with Lion's Mane mycelium.

Strain Isolation

Whether you use a spore print or a spore syringe, you'll be left with multiple strains growing from the inoculated portions of your dish. Depending on the species, they should grow out on your dish in a few days or a few weeks. But after you get 1 to 2 inches (2.5 to 5.1 cm) of growth from your inoculation point, you should be able to see different types of growth emanating from the starting position. These are different strains of your given species, and the mycelium of these different strains often look different from one another. Your goal will now be to isolate the most vigorous and healthy strains growing out on your dishes by transferring a small portion of them to a new dish. This process is called *strain isolation.*

Each of the different strains isolated on the dish not only has mycelium that looks different, but also has many other characteristics that differ from its neighbors. It may have different rates of mycelial or mushroom growth, or it may ultimately produce mushrooms that look different from other strains. One strain may produce more but smaller mushrooms. Another may produce fewer but larger mushrooms. Some may grow better on one substrate than on another. Some may produce mushrooms that taste better, or look better, or produce a different number of spores than another strain.

The goal of strain isolation is to get a single strain growing from your petri dish. Many species will exhibit rhizomorphic growth as seen above.

The point is that each individual mycelial strain produced by inoculating agar with spores will have slightly different characteristics than all the others on the dish.

Strain Development

Strain isolation allows you to capture and test the genetics of several individual strains, and to determine which work best with the method you're using to grow. Here are some characteristics to look for as you make these determinations.

Fast Growth

First, look for strains that grow quickly. You won't want to isolate slow-growing strains, as one of the primary goals of cultivation is to grow mushrooms as quickly as possible. There should be a number of aggressive strains growing out from your inoculation point, outgrowing and outcompeting other strains.

Another growth-related characteristic is adaptability. After you make your first isolated transfer, you want a strain with a short recovery time, which means one that will adapt quickly to its new media and begin growing. You also want the strain that quickly colonizes grain jars once that transfer occurs, quickly colonizes the fruiting substrate, quickly starts to pin, and quickly grows into fully formed fruitbodies. While the initial growth rate in a petri dish can be a good indicator of future growth, take notes about the individual strain throughout the entire growth process to ensure that it will meet all your needs in the future.

Temperature can affect the rate of growth of specific strains, even within a given species. A common species like Shiitake can have strains that have been isolated for cold-weather fruiting and other strains that fruit better in warm weather. Look for a strain that grows and fruits best in the temperature that's ideal for your environment.

Fruits Properly

Strain development doesn't stop with the mycelial phase; you should continue to develop a strain throughout the entire fruiting process. Although you want to find a strain that begins fruiting fast, you also want to evaluate how well the strain fruits. The best strains produce a very large pinset on your substrate, with a large portion of the initial primordial forming fully into fruitbodies. Some strains will have a higher percentage of aborts than others. Once they're fully formed, you want to ensure that the fruitbodies look good, meaning that they're free from malformed or mutant mushrooms, look and smell as they're supposed to, and have a good texture. You want to ensure that your substrate produces great second and third flushes as well, and that the time between flushes is not too great.

All these characteristics are fairly easy to evaluate, and will help you develop great strains that work well for you long into the future. The development of great strains takes time, patience, and luck, but it's well worth the effort.

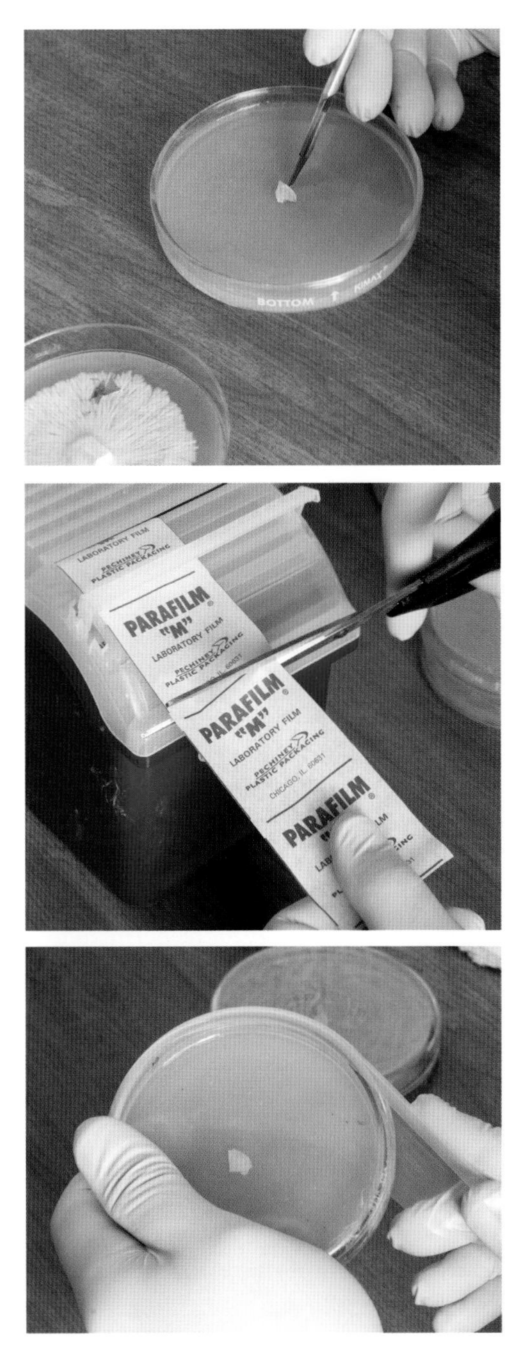

After a dish is inoculated, Parafilm is commonly used to seal the petri dish. This helps the dish to retain moisture and prevents contamination.

Performing an Isolation

To perform a strain isolation, you should have a petri dish on which spores have grown out mushroom mycelium. As the mushroom mycelium grows out, you'll begin to see multiple areas of growth differentiating themselves on the dish. These are all different strains of your culture that have resulted from individual spores mating. Cultivators refer to these individualized areas as "sectors" of the dish. Once this sectoring begins to occur, your goal is to transfer small pieces from the edges of these individual sectors to new petri dishes. Although your initial dish may have many sectors, after you make a transfer to a new dish and the new dish begins to grow out, there should be far fewer different sectors visible. This process isolates individual strains of your species, with the goal of testing the growth habits of these individualized strains.

Numbering

If you transfer between five and ten sectors of your initial dish to new dishes, you'll have a variety of genetics to test and eventually choose as a final strain to save and cultivate. It's best to come up with a numbering system to help you keep track of each strain, as well as the generation.

For the first generation from spores, you might start by calling the dish something like LE-0 for Shiitake. The scientific name for Shiitake is *Lentinula edodes*. The "0" denotes a culture started from spores.

As you begin to see sectoring on your dish, you'll isolate multiple strains from your dish. Give each transfer a new number to differentiate the individual strains. Thus, after the first round of transfers, label the edge of each new dish LE-1, LE-2, LE-3, etc., to represent each isolated Shiitake strain.

It's also important to keep track of each dish's generation. Paul Stamets invented a nomenclature called the "P Value system" that aids in this process. Your first transfer away from spores can be thought of as the first generation, and given the label P1. Once that dish is grown and transferred again, the new dish is labeled P2, and can be thought of as the second generation away from the initial culture.

The goal is to isolate a single strain of your given species in as few transfers as possible. This keeps the genetics of the culture as fresh as possible. With each successive transfer, the genetic quality of your culture will degrade. The culture grows by replicating its genetic material and undergoing cell divisions. By maintaining cultures with the lowest P Value possible, you'll stay very close to the original genetics, and your number of cell divisions will be very low. This is the ideal situation for maintaining the strongest, most aggressive cultures long into the future.

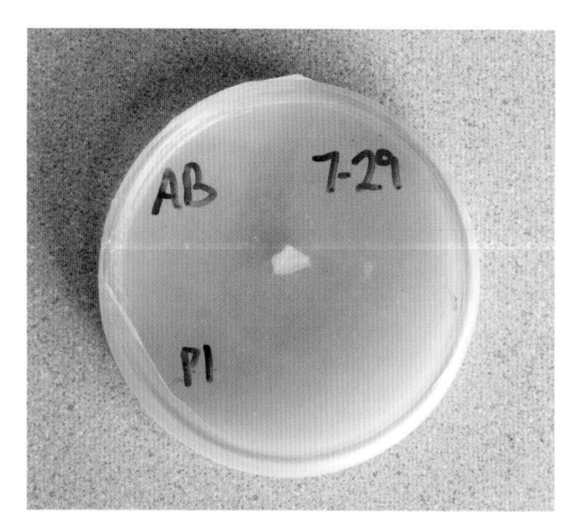

Petri dishes are commonly labeled around the outer perimeter. "AB" stands for *Agaricus bisporus*, the White Button mushroom.

Transfers

The ideal location for transfers is from the outer edge of your growing culture. This area, at the edge of the expanding hyphae, is where the largest numbers of cell divisions happen and growth is strongest. Don't allow your culture to grow entirely to the edge of the dish before making your transfer. Ideally, you should keep the number of cell divisions as low as possible within a given dish. You'll accomplish this by making your transfers as soon as you're clearly able to identify the growing sectors. You should also be aware that contaminants can find their way over the edge of the dish and onto the perimeter of the agar medium. By making your transfer before the mycelial culture reaches the edge, you'll eliminate this potential contamination.

As a home grower, you may not have the time to work in your lab on a daily basis, so you may not be able to make your transfers before your growing culture reaches the edge of the dish. Although transferring your culture before it reaches the edge of the dish is optimum, it's not absolutely necessary for good growth after a transfer.

When starting from spores, you may have to transfer your strain to a new dish three or four times before you're able to isolate a single strain without any visible sectors. This is a normal part of the process that any grower must work through.

Sourcing from a Culture Bank

Strain isolation is necessary only if you're starting a culture from spores that have multiple strains of the species on the same dish. Another option is to buy cultures that have already been isolated and are geared for the commercial production of mushrooms. There are several culture banks around the country that specialize in providing living cultures to researchers and growers. One of the best resources is the culture bank from the Penn State Department of Plant Pathology and Environmental Microbiology. (See Resources, page 228.) Cultures of mushroom species from all over the world, many of which are used for commercial mushroom production are sold here. This means that the strains will usually be fast-growing and high-yielding. Species from Penn State's collection are expensive, but you'll save an incredible amount of time as compared to isolating your own strain. You'll also be getting a strain that has been developed and isolated by professional mycologists.

Cultures from a culture bank usually arrive in a petri dish or a culture slant. When they arrive, simply cut away a small section from the dish and place it in the center of a fresh petri dish, just as if you were performing a strain isolation. There is no need to do multiple transfers as in isolation unless there is contamination that presents itself on the dish when you perform a transfer.

Sourcing from a Mycology Store

There are many online sources of mushroom cultures. The quality of the culture (e.g., the number of generations from the source culture, how it has been developed, how it has been stored, and how well it ultimately fruits) may vary significantly among sources. Consider the reputation of the company before you buy.

When you buy a culture from a mycology supplier, you'll usually receive a petri dish or a culture slant. These can usually be transferred directly to a quart of grain from the dish you receive, but it's best to make transfers to several of your own dishes before advancing your culture to grain. This is primarily to make sure the dish you receive has not picked up any contamination along the way, and that you're starting with a clean culture.

Liquid culture syringes may also be useful when considering agar work. Most mushroom suppliers sell liquid culture syringes for significantly less than petri dish cultures. The advantage is that you can easily make your own petri dish from a drop of two of the liquid culture syringe. Before you make a dish, be sure to eject a few drops of the culture from the syringe before sterilizing the needle; if any contamination made its way around the tip of the needle, it will be ejected before you begin your work. Also remember to start your flame sterilization at the plunger end of the needle and work your way up to the tip. Once you sterilize the needle, eject a drop or two of the liquid culture into the center of the plate. Liquid culture syringes contain a single isolate, so there's no need to perform a strain isolation.

Dealing with Contamination

One of the main advantages of working with agar is that contamination is not as much of a concern as when you're working with most of the other methods. If contamination appears on your plate, you can isolate the culture away from the contamination, just as if you were performing a strain isolation. Growers usually refer to this as "cleaning up a culture."

Fungi

The most common contaminants in agar cultures are types of fungi. These common household molds usually start as a small white dot somewhere on the surface of the plate that quickly turns green, blue-green, black, gray, or red. If the contaminant appears to be near the inoculation site, could be a result of contamination in the culture you were previously transferring from or contamination in your instruments. Both of these potential sources of contamination need to be checked before you make additional plates.

If the contaminated area is not near your inoculation site, check other potential sources of contamination, such as your sterilization methods (pressure cooking time, cool-down method, and location), the methods you use to pour the plates, and the effectiveness of your flow hood or glovebox. If the spots of contamination appear on many plates or in any significant quantity, take a hard look at all of these potential areas of contamination. If you have only a spot or two of mold on a couple of plates, it's much less of a concern, but it would still be wise to think through all of your processes. Even working in front of a 99.99% efficiency HEPA, you'll see a small amount of contamination from time to time, so mold is not always indicative of larger sterility issues.

Bacteria

Bacterial contamination is somewhat tougher to clean up than mold. While molds present themselves as a mycelial-like culture that soon turns a different color, bacterial contamination tends to present itself as a watery or milky-looking substance on the surface of the agar. It will not grow out to cover a plate as fast as a mold colony will, but it can still cause problems later if it's not properly identified.

Antibiotics can be used in your agar formula to help control bacterial contamination. Gentamycin sulfate is most commonly used because it can be added to the agar mixture before sterilization. Many of the other antibiotics must be added after sterilization, because their chemical structures degrade at high temperatures.

The recommended amount is 50 to 100 mg of Gentamycin per 1 liter (1 qt.)

of your agar solution. Adding more than this will not offer any additional benefit. Stamets recommends 100 mg per liter based on Gentamycin that's 60 to 80 percent pure, so you may need to adjust your formula based on the purity of the product you obtain. Use less (closer to 50 mg) of a product that has a higher purity.

You may be able to find this antibiotic at farm and feed stores or with the help of your local veterinarian. Otherwise, many of the mushroom supply stores online can help you acquire it. It comes in powder or liquid form.

Sealing Dishes

Whenever you make a new dish, it's important to seal it to maintain moisture levels and prevent contamination. A plastic sealing product called Parafilm, which has an extraordinary ability to stretch, is commonly used for this process. Without a barrier, moisture will wick away from your agar over time and your culture will dry up in the dish. Because most home cultivators don't invest in cleanrooms, a plastic sealing material also helps keep the dish sealed against contamination.

Bacteria and fungi are common contaminants on agar. Sealing dishes with Parafilm will help prevent contamination.

Cloning

Spores and living mycelia are not the only way to start a mushroom culture. Literally, any section of a fully grown mushroom fruitbody can be used to start a mycelial culture. Whenever growers use a section of a mushroom to start a new culture, they create a cloned culture.

A clone is an exact genetic replica of a living fruitbody. Isolating a viable, high-fruiting culture from spores requires a significant amount of time and resources. By choosing to start a culture from a living mushroom, the grower can skip most of the strain isolation steps and be left with a clone of a single, high-fruiting strain of a given species. Cloning is probably the easiest way to attain a high-quality culture of edible species.

Finding Source Material

Grocery stores or farmers' markets are probably the best places to look for common edible species to clone. Any mushroom available for commercial purchase can be cloned on agar by any cultivator. The main advantage of obtaining mushrooms from these sources is that you'll be cloning a species that has already been deemed good enough for commercial cultivation. The mushroom you're cloning will be from a fast-fruiting, high-yielding strain of that species. Some people claim that these mushrooms are hard to clone, or that too many cell divisions have occurred to make the culture viable, but I question those sentiments.

As a home cultivator, you don't have access to the expensive lab equipment and personnel that would allow you to attain the best cultures possible for cultivation. But you don't need it, because you have access to some of the best cultures available at your local grocery. The genetics might not allow you to continue cloning the culture indefinitely, but you have a good source to go back to for a new initial culture — the same grocery store where you purchased the source culture.

The Wild

Some growers prefer to find fruitbodies in the wild to clone. There are several considerations if this is your goal. The first

Use Antibiotics Sparingly

The vast majority of agar work, including cloning, should not require the regular use of antibiotics. They should be used sparingly, such as when you need to transfer from a plate that has bacterial contamination, or if you're starting a wild tissue culture. If you find that bacteria or other contaminants are a continual problem, examine your sterilization procedures before considering the use of antibiotics.

is proper identification of the species you collect. There are thousands of different mushroom species, and many of them are difficult to identify, even for seasoned mycologists. Don't go out into the wild and start cloning species unless you're very familiar with mushroom taxonomy. If you're new to identifying mushrooms in the wild, find a local mycological association where seasoned identifiers can teach you the skill.

The second consideration is whether the wild species you're considering can be cultivated. Mushrooms can grow from a variety of different substrates, but the most common species for home growers are wood-loving mushrooms such as Shiitake, Maitake, Lion's Mane, and Reishi. These and most of the other common edibles are grown from sawdust, and are found on wood in the wild.

One specific group of edible mushrooms that are notoriously difficult to cultivate are mycorrhizal mushrooms. In order to grow, these mushrooms develop a symbiotic relationship with the roots of trees. They get their carbon from the photosynthetic activity in the leaves of trees, and they provide nutrients from the soil back to the tree. This is a complicated symbiosis that is, so far, impossible to replicate in a controlled environment. Chanterelle, Bolete, Morel, Milk Cap, and many other species are mycorrhizal. Research is ongoing on this symbiosis, but many of these species are impossible to culture in a petri dish, much less grow into a fully formed fruitbody.

The final consideration with cloning wild species is how well the species will do under controlled conditions. Some commonly cultivated species, such as Maitake, don't grow very well from wild isolates. It may take a long time to develop a strain of the species that can be grown successfully under controlled conditions.

Selecting the Part of the Mushroom to Clone

Before learning how to clone a mushroom, it's best to understand the goal. A clone can be taken from any tissue of any mushroom that can be cultured, but some sections of the mushroom yield better results than others. For example, any tissue that has been exposed to the open air is not ideal. This increases the likelihood that the mushroom has picked up mold spores or bacteria. To avoid this, it's best to use tissue from the interior of the mushroom.

When I first began cloning, I'd slice the mushroom open lengthwise using a scalpel to access the interior. This can be done, but be aware that the scalpel may pick up contaminants from the outer surface of the mushroom and drag them into the interior surface. Your goal is to remove as many potential sources of contamination as possible, and this is one of them. Instead of cutting into the mushroom to expose the inner flesh, simply rip the mushroom open with your hands. Most edible mushrooms will rip open fairly easily, and the inner surface will never have come in contact with objects that may be harboring contaminants.

Cloning Method

Materials

- Fresh fruitbody
- Tweezers
- Scalpel
- Petri dish with agar
- Latex gloves
- Shot glass or half-pint canning jar (optional)
- 3% hydrogen peroxide (optional)

Begin by sterilizing your materials. This includes your agar formula and containers (unless you're using plastic dishes that have already been sterilized). Wrap metal tweezers and your scalpel in foil to cook along with the agar.

Prepare your work surface or glovebox. Set the mushrooms you want to clone on a piece of paper towel. Set your wrapped utensils on another piece of paper towel. When your agar plates have cooled down and solidified, put on your gloves and unwrap your tweezers.

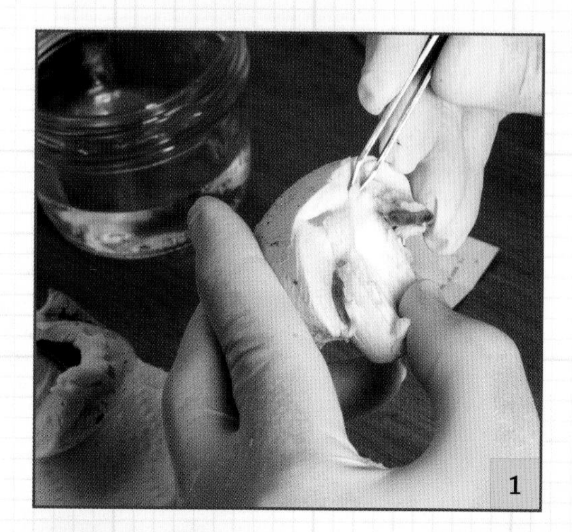
1

1 **Open the mushroom.** Rip the mushroom you're cloning in half, exposing a section of the mushroom tissue that has not previously been exposed to unsterile air. Remove a small section of mushroom tissue, about the size of a grain of rice or rye, from this sterile section.

2 **Dip the flesh into a jar containing a 10:1 solution of water to hydrogen peroxide.** This is an optional procedure that will help prevent bacterial contamination.

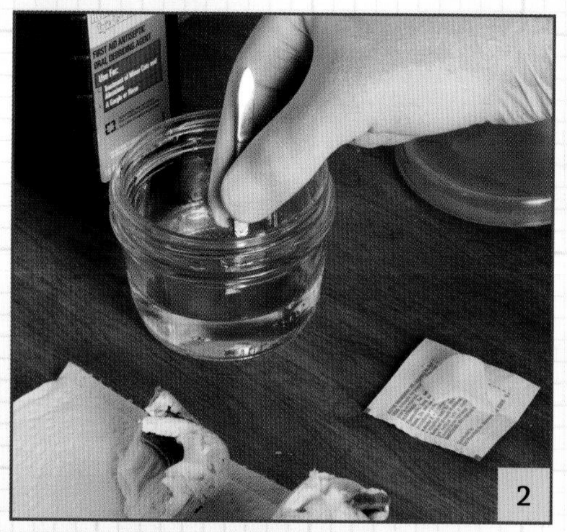
2

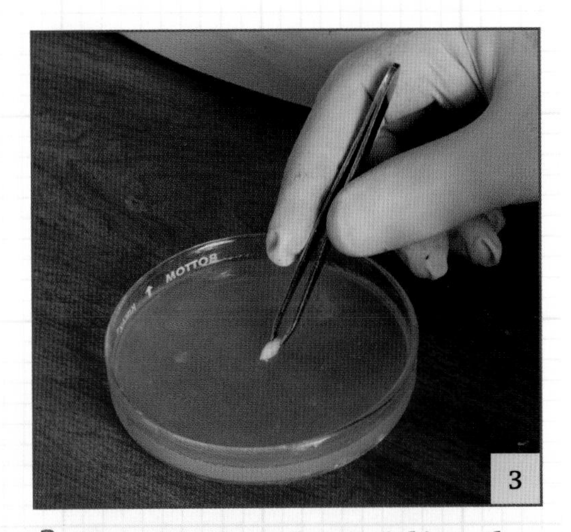

3 **Place this piece of material into the center of the petri dish.** I usually pierce the surface of the agar with the tweezers, and force the tissue down into the agar medium. This gives the tissue much more surface contact with the agar than if you just set it on top of the medium. The culture will still grow out from the center of the dish, as with any other transfer. Place the lid on the dish. Heat sterilize your tweezers before obtaining another piece of tissue for the next dish. It's usually best to clone 5 to 10 dishes of each species you're working with. Seal your dishes with Parafilm once all transfers are complete.

Label your dishes, noting the species, that it's an initial clone culture, and the date. I also label the dish or make a separate note of the geographic location of the source material.

For most gilled mushrooms, it's best to open up the entire mushroom vertically, through the center of the cap and stem. This will expose the most tissue, allowing you multiple locations to obtain clone material from.

The optimum location to extract tissue from gilled mushrooms is where the tissue from the stem encounters the "context" of the cap. The context is the portion of the cap above the gills, but below the upper surface of the cap. It's usually a spongy, soft tissue, easily removed with tweezers. You can also use tissue from the inner surface of the stem. It's generally more fibrous than tissue from the context, so you may need to use a scalpel to remove a section.

Do not try to obtain clone material from the gills of a gilled mushroom. The gills are where the spores are produced, so if you try to clone the gill tissue, you're likely to introduce spores into your culture as well. This will essentially defeat the purpose of cloning, as you'll be offering other cultures the opportunity to form along with your clone.

For mushrooms without gills or oddly shaped mushrooms, you may have to get a little more creative to find sterile tissue. Maitake is a good example of an oddly shaped mushroom. Maitake's rosette petals are very thin, and it's very difficult to obtain sterile tissue from most of the fruitbody. For mushrooms with this shape, it's best to get tissue near the base of the mushroom. The base of this species is fairly solid and should have a lot of useable tissue.

Common Problems with Cloning

One of the most common problems with cloning is bacteria, especially from wild specimens. Cloning fresh mushrooms from your own grows or from grocery stores should not bring about significant problems with bacteria, but mushrooms cloned from the wild will. If you're cloning wild mushrooms, please refer to some of the antibiotics previously mentioned as an addendum to your agar formula. They will simplify your attempts to attain a clean clone from the wild.

There are a few other procedures that a grower can use to minimize bacteria when cloning a wild specimen. The first is to soak or rinse off the fruitbody in something that will kill off contaminants, such as iodine or hydrogen peroxide. Some growers wash off the exterior of the fruit-body with iodine, and also dip the clone tissue into iodine before placing it on the petri dish. Others use 3-percent hydrogen peroxide in much the same way. Either of these will help clean off wild specimens before you begin the cloning procedure. You should still take tissue only from the mushroom's unexposed interior surface, regardless of whether you use either of these soaks or washes.

Saving Your Strains

No matter which procedure in this chapter you use, the end result is a culture of a single strain of a species that you want to retain for the future. A culture in a petri dish can easily last a year or more if it's refrigerated. Cultures that are properly prepared for storage can last many times that, so you will have access to that same viable culture long into the future.

Most growers save mushroom cultures in test tubes using slants. Slants are created by pouring agar into a test tube and letting it solidify at an angle. This angled surface allows for greater fungal growth surface area in the tube than if the agar is allowed to solidify upright. Some test tube racks are designed to sit at specific angles while the test tubes cool. One of these racks is worth purchasing if you're saving strains.

To inoculate the test tubes, simply cut a small piece of culture from your petri dish and place it on the agar in the tube. Let the agar grow out for several days to completely colonize the surface in the tube. Finally, fill the tube with sterilized mineral oil so that the entire agar surface is submerged in the oil. This oil will prevent the culture from drying out, and, when combined with refrigeration, will allow the culture to stay viable for many years into the future. You can also use liquid paraffin instead of mineral oil. Either of these oil overlays should be sterilized at 15 psi for 1.5 to 2 hours and allowed to cool before being added as a top layer to the test tubes.

Once you're ready to reuse this culture, simply pour off the mineral oil, cut away a section of the culture, and inoculate your new dish.

CHAPTER 12

LARGE-SCALE GRAIN SPAWN

If you want to continue expanding your mycelial cultures, your next step is to learn to produce larger quantities of grain spawn. Grain spawn, generated in polypropylene spawn bags, allows you to turn one quart jar of colonized grain into 5 or 6 pounds (2.3 to 2.7 kg) of colonized grain; however, it is somewhat more difficult to work successfully with these larger quantities than with quarts of grain.

Small home growers will generally not need the amounts of substrate produced using spawn bags. You should consider mastering this process only if you regularly need more than 20 quart jars at a time for your operation. When working with less than 20 jars, I actually prefer to work with quarts. Smaller units allow you to identify potential contaminants more easily and isolate them more effectively. Grain spawn in bags also take significantly longer to pressure cook than quarts.

Spawn Bags

Spawn bags used in mushroom cultivation are often called "filter patch bags" because of the filter built into each bag. Without the filter, carbon dioxide would begin to build up within the bag and inhibit growth. Order spawn bags online from a mushroom supply store. There are no adequate substitutes available elsewhere. For more information on spawn bags, see page 156.

Size

Most mushroom supply stores sell spawn bags in two sizes: medium and large. Both are about the same height, but the large bags are a bit wider at the base. A medium spawn bag will hold about three level quart jars of grain, while the large bags hold about six level quart jars. Some supply shops carry slightly smaller and larger sizes as well, but for most purposes, I'd stay away from them. A better substitute for the smaller bags would be quart jars, as they're reusable and more durable. Bag sizes larger than 6 quarts require even longer sterilization times and can be difficult to sterilize effectively.

Thickness

Spawn bags are available in two types: standard or thick. Standard spawn bags are usually 2.2 to 2.5 mil thick, while bags labeled "thick" are generally 3.0 mil. I have used both of these styles successfully for grain spawn.

Filter Types

The difference between spawn bag filters is their effectiveness — the size of the particles each filter allows through. The most commonly sold ratings are 0.2-micron, 0.5-micron, and 5.0-micron filters. For grain spawn, you will want 0.2- or 0.5-micron filters. They're efficient enough to screen out most potential contaminants.

Contamination

The primary reason growers fail with grain spawn is contamination. Larger amounts of grain mean more potential problems with preparation and contamination for growers, especially new growers.

Experienced growers can identify contaminated quarts of grain much earlier than new growers. Contaminants are not always readily visible. For example, a small culture of mold or bacteria may begin to grow in a jar, but get overtaken by the mushroom mycelium you're trying to propagate; however, this doesn't kill the detrimental colony, it's just hidden within the mycelial mass. If you try to break up and transfer a jar with a hidden colony of mold or bacteria, your subsequent cultivation processes aren't likely to be successful. This is especially true with bags of grain. The ability to identify misfit cultures is one of the main reasons that experienced growers have more success with these advanced techniques. They're able to spot contamination issues early in the process, before they have a chance to cause much greater harm.

Another opportunity for contamination when using spawn bags occurs during the sterilization process. Large quantities of grain require significantly longer sterilization times than do quart jars. All the grain in the bag, including the grain at the very center of the bag, must reach the proper temperature for the proper amount of time. If even one small section of the contents of the bag isn't sufficiently sterilized, the entire bag is likely to go bad before it's fully colonized.

A final contamination issue to address is the problem of endospores, which allow certain types of bacteria to survive very high temperatures for extended periods of time. If you're having problems with bacteria contaminating your grain, endospores may be the cause, but there's also a simple fix. Hydrating your grain using the soaking method will encourage endospores to germinate before the sterilization cycle starts. Once endospores have germinated, they can no longer withstand grain sterilization temperatures. Soaking your grain before pressure cooking ensures that all bacteria will be killed during sterilization.

Grain Spawn Methods

You have many different grain options for grain spawn bags. Rye, popcorn, and millet (wild birdseed) are the most common options. (See chapter 6 for more information about preparing and hydrating grain.) The best grain to use depends entirely on what you plan to do with the spawn once it's colonized. Most of the time, you'll be spawning the grain to another substrate such as straw, manure, or compost, so it may be best to choose a grain with small kernels that will give you many inoculation points. If you're making a straight grain casing, the number of inoculation points is less of a concern, so popcorn may be a better choice.

Large amounts of grain spawn can be produced with wild birdseed, rye, popcorn, or (as shown above) a 50/50 mixture of popcorn and birdseed.

I rarely recommend elaborate mixtures of ingredients in mushroom cultivation, but grain spawn is one area where I make a mixture from time to time. Using a 50/50 mixture by volume of wild birdseed and popcorn offers the high number of inoculation points associated with wild birdseed, but also the faster colonization times associated with popcorn. Using this simple mixture rather than a bag of wild birdseed or millet can save 2 or 3 days of colonization time, and is worth considering if you're on a tight schedule. Hydrate and rinse each grain individually. Place them together in the spawn bag for the final sterilization cycle.

Wild Birdseed in Bags

My top choice grain for creating grain spawn at home is wild birdseed. You'll be hydrating and sterilizing a significant amount of grain, and birdseed is both readily available and inexpensive. Always look for seed that has millet as the primary ingredient.

For more information on wild birdseed preparation, see page 124. Here are some tips for preparing larger amounts:

First, make sure you wash the grain off thoroughly after it has been hydrated and before it's loaded into bags. If you don't wash it off well, starches will accumulate after cooking and individual kernels may stick together to form a brick that's hard to break up. Starchy grain can also make the bag slightly slimy, and the mycelium doesn't always thrive in slimy conditions.

Second, dedicate a small pitcher to the process of loading the bags. If you know the weight of your pitcher when filled level, the process of filling bags can be quick and effective. For example, say that two level pitchers of grain equal the required weight. Scoop twice and the bag is done. You could use permanent marker to mark a fill line if a level fill doesn't provide an appropriate weight.

I normally use 5 to 6 pounds (2.3 to 2.7 kg) of hydrated grain for a large spawn bag; that's roughly six level quart jars. If you use much more, it will be difficult to shake the bag; if you use much less, you're not making the most effective use of your bags.

The total amount of wild birdseed you'll need depends on how much your type of seed swells as it's being hydrated. Different brands and mixtures will swell different amounts, but a 40-pound (18 kg) bag of dry seed should make about seven large spawn bags of hydrated seed. If you're making 50/50 wild birdseed/popcorn bags, a 40-pound (18 kg) bag of dry wild birdseed should make about 15 large spawn bags (using a similar volume of corn as well).

Popcorn in Bags

As with wild birdseed, the first step is to hydrate your popcorn. I prefer the pressure-cooking hydration method (see page 122). In just a couple of hours, it will give you hydrated popcorn that is ready to be sterilized. How much popcorn you need per bag depends on several variables, including how long you cook the popcorn, at what pressure level, the level of water in the cooker, how long it takes to cool down, and so on. The individual kernels swell to different sizes based on all of these variables. Overall, you should prepare 3 to 4 pounds (1.3 to 1.8 kg) of dry popcorn kernels per bag of spawn you plan to create.

Remember that to help remove additional starch during the hydration phase, you should take the corn out of the water in the cooker just after the pressure is down, but while the cooker and water are still hot. This will leave most of the starch suspended in the hot water, and it can be easily poured off. As with wild birdseed, it's imperative that you thoroughly wash the corn after it's been hydrated. This will remove any additional starches and keep the corn from becoming sticky and solidifying after the sterilization cycle.

Corn can be hydrated by pressure cooking; be sure to clothespin the top of the spawn bag before putting it into the pressure cooker.

Working with Bags

Using spawn bags requires some practice. These tips may help make your first experiences more successful.

Before Sterilization

Before you load your spawn bags into the pressure cooker, they must be folded so the contents don't fall out. Bags shouldn't be completely sealed or they might burst. The best way to seal the bag for cooking is to fold the top part of the bag over, pushing all the air out. Then fold the top over two or three times as pictured. Finally, place two or three clothespins over the fold to hold it in place. Common wooden clothespins will withstand many runs through the pressure cooker.

Loading Your Cooker

When cooking spawn bags, you'll need a 20-quart (20 L) cooker or larger. Mirro, Presto, and All American cookers in the 20 quart (20 L) range will hold three large spawn bags filled with grain. Larger cookers, such as the All American 941 (41 quarts (or about 41 L) will hold six large spawn bags of grain.

When pressure cooking spawn bags, it's vital to ensure that you have enough water in the cooker to last for the entire sterilization time. Even if you use a standoff plate, spawn bags may not allow enough water to be placed in the bottom. I've had cookers run dry when cooking bags, and the bags usually melt and the grains actually cook, making them unusable. To fix this problem, place two layers of ring bands (or something similar) under the standoff plate, to allow space for more water in the bottom of the cooker.

Cooking

There are several issues you are likely to encounter when pressure cooking bags. The first is determining the correct sterilization times. Large spawn bags with 5 to 6 pounds (2.3 to 2.7 kg) of grain will need to be pressure cooked at 15 psi for 2.5 hours to ensure the grain at the center of the bag has reached the proper temperature. I have talked with numerous people who were having problems with bacterial contamination, and were pressure cooking for 2 hours. Once they increased the cooking time by that additional ½ hour, their problems were solved.

The second issue is preventing spawn bags from melting in the cooker. Some growers find that their bags melt or become stretched during the sterilization cycle, particularly in 20-quart (20 L) cookers that are heated on the stove. If cooked properly, spawn bags should come out of the cooker roughly the same consistency and thickness that they went in.

There are several ways to prevent melting bags. The first is to heat the cooker up more slowly. These poly bags are meant to hold up to sterilization temperatures, but they don't tolerate quick temperature changes very well. If you heat the cooker up too quickly, it may cause the bag to fail. The same holds true during cool-down: if you release the pressure during cool-down, the quick change in pressure and temperature can also weaken the bag. Failed bags could also be an indication that you're letting the pressure cooker reach temperatures or pressures that are too high. If your cooker has a pressure gauge, try not to let it go higher than 15 psi. If you don't have a pressure gauge, ensure that your jiggler is not constantly jiggling throughout the sterilization cycle.

If none of these suggestions solves the problem, consider wrapping a hand towel around the bags inside the pressure cooker so they're not touching the sides. This will allow less heat to be transferred to the plastic from the metal sides.

Transferring Jars

After your bags of grain have been sterilized, it's best to wait until your cooker is completely cool before transferring jars of colonized grain into the bags. Allow the cooker to sit overnight. When you're ready to begin transferring jars, fire up your flow hood and let it run for an hour before you begin (see chapter 5). With your hood running, break up one fully colonized quart jar per spawn bag you prepared (see chapter 6). For medium bags, use one quart (1 L) jar of spawn per two bags, or ½ quart (500 mL) per bag. You shouldn't need anything else in front of the hood except an impulse sealer (see page 157). Simply pour the contents of your jar into the grain bag and seal it.

Incubating

I like to lay spawn bags flat for incubation, as they tend to colonize more efficiently that way. It should normally take less than 2 weeks for the bags to become completely colonized.

Given the larger amounts of spawn in bags, I don't recommend putting the bags in an incubator. Much more heat will be generated from a bag than from a jar, so keep the bags at room temperature for incubation, and don't stack them on top of each other. The heat will have nowhere to go and will slow the mycelial growth.

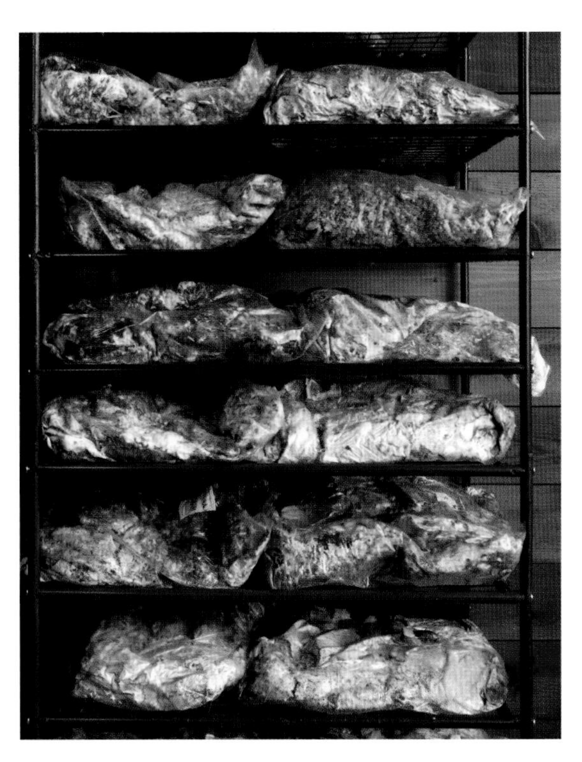

Spawn bags colonize more efficiently when they're laid flat to incubate.

FAQ for Large-Scale Grain Spawn

How long can I keep my bags in a cooker before inoculation?
Because growing mushrooms is a race between the mycelium and contaminants, you'll want to inoculate soon after the bags are cool. That being said, common time constraints often mean they sit in the cooker for a day or so before inoculation. Try not to let them sit longer than a couple of days.

How long will my colonized spawn bags last?
As with any type of spawn, use it as soon as possible after it's fully colonized. That being said, it will still work fine after 1 or 2 months, assuming no contaminants have set in. The longer you wait, the more likely there will be contamination.

I always have problems with bacteria in bags. Is there anything I can do to avoid this?
If you've tried soaking your grain for hydration, and then sterilizing it for longer periods to minimize the presence of endospores in the grain, and you're still having problems, there's one other thing to try: reduce the moisture in your grain. This involves shortening the soak time or pressure cooking time of the grains during hydration. Sometimes excessive moisture in the bags or in the grain encourages bacterial growth.

Will a vacuum sealer work as an impulse sealer?
Yes, although you don't want to use the vacuum function. Grain bags need air in them to help the mycelia colonize, and the air in the bag allows room for the bag to be shaken.

Some bags have injector sites. Should I inoculate directly into a spawn bag?
I would say no. Injecting with spores or a liquid culture will take a while to colonize and eat up a lot of your fluid, especially for the large spawn bags. I would suggest making grain jars first, and then doing a G2G transfer.

Can I use a glovebox with spawn bags?
Yes, but I don't recommend it. Spawn bags take up a lot of volume, so if you're working with them, you should have progressed to using a flow hood. Gloveboxes are too small and difficult to work with for these procedures.

CHAPTER 13

BULK SUBSTRATES

Bulk substrates are large amounts of any nutritious medium used to grow mushrooms. The term commonly refers to fruiting substrates other than sawdust. Sawdust is technically the most commonly used bulk substrate, but most cultivators who use this term are referring to other media such as straw, manure, or manure-based compost.

Oyster mushrooms, for example, fruit well on sawdust, but most people choose to fruit them on straw because it doesn't have to be sterilized before spawn is applied. Some Agaricus species, such as Portabellas, grow best on manure-based substrates, which don't require sterilization either. Instead, straw and manure-based substrates rely on pasteurization to make them suitable for mushroom cultivation. Pasteurizing substrates saves a significant amount of time and expense when you're working with large amounts, especially if you're a home grower without access to large commercial pressure cookers.

Pasteurization is a process similar to sterilization, but it requires lower temperatures. Sterilization occurs by attaining 250°F (121°C) for 20 minutes; when pasteurizing, you usually bring the substrate to between 145 and 165°F (63 and 74°C) for 1 hour. The simplest way to explain pasteurization is to say that it kills off the "bad" bacteria and contaminants, and allows the "beneficial" bacteria to survive.

This chapter explains some of the easiest ways to pasteurize small and large amounts of bulk substrates and prepare them for use.

Straw

No special type of straw is required for mushroom cultivation; a bale from any garden center or hardware store will do. Smaller amounts are available at hobby stores that carry mini-bales for craft projects, and can work well if you don't have room to store an entire bale.

Cutting Straw

Straw used as a substrate should ideally be cut into 2- to 3-inch (5 to 8 cm) pieces before being hydrated and pasteurized. Although cutting up the straw is not absolutely required, smaller pieces can be packed more tightly into the container you'll be using, and this will help shorten colonization times. Cut-up straw has more exposed surface area and can potentially offer more nutrients to the mycelium. Also, short pieces of straw are easier to work with.

There are several ways to cut straw, some more labor intensive than others. If you don't mind a bit of labor, and only need a small amount of straw, you can cut it up, clump by clump, using a large pair of garden shears. You'll have a full pillowcase in 10 or 15 minutes, and it's a good choice if you live in an apartment without much room or outdoor working space.

You can also put straw into a garbage can and cut it up with a weed eater. Just cut a slit in the lid, insert the neck of the weed eater, and start chopping like a madman. Some people run the straw through a wood chipper. Both of these methods can process a lot more straw but also require outdoor working space.

Straw that's to be used as a substrate should be cut into small pieces. Large garden shears are fine for cutting up small quantities of straw.

A string trimmer can make quick work of chopping straw. Simply cut a hole in the lid of a clean trash can so that the straw stays contained while the trimmer is working.

Hydrating Straw

Like all mushroom substrates, straw must attain the proper moisture level before it can host a growing culture. To ensure the correct moisture level, you should hydrate it before proceeding to pasteurization. You can either soak your straw in a hot-water bath or steam it. The hot-water bath methods in this chapter will hydrate the straw during the process much more than a steam treatment, and thus require shorter hydration times. If you're using a hot-water bath, soak the straw for 1 to 2 hours before putting it in the bath. If you're steaming the straw, soak it for 12 to 24 hours before implementing the steam cycle. With both of these methods, let all the excess water from the soak drain off before proceeding further.

Manure

The idea of working with manure may sound gross, but let me reassure you, you'll never work with anything that has the smell or texture of fresh manure. In fact, if the manure smells bad or has a disgusting texture, it's too fresh for use in mushroom cultivation.

Mushrooms require manures that are well composted, with the smell and texture of dry dirt. These "aged" manures, whether from horses or cows, have been sitting out in a field or behind a stable for quite a while. Horse and cow manure are basically interchangeable for any mushroom species that grow from this substrate. If you live in the country, anyone near you who has horses or stables probably has more manure than they'll ever use, so don't feel shy about asking for some. You can also check newspapers in spring or look for ads on Craigslist throughout the year. Garden centers sometimes carry locally sourced, composted manure.

Horse manure should be well aged before it's used as a substrate.

Most garden centers carry branded, bagged manure as well. I've had mixed results with these products, so you may have to do a bit of testing before you find a brand that works well with the species you want to grow. One way to improve the efficacy of bagged manure from garden centers is to mix it 50/50 with another substrate, such as vermiculite.

Crumbling Manure

Before it can be used as a substrate, manure must be broken into individual pieces that are smaller than a marble. This is possible only when manure is dry, so any type of manure except the bagged varieties must be dried for several days. Spread your manure out on newspapers on the porch or in the yard in dry weather, and let the sun dry it. After the manure has dried out enough to crumble easily, it's ready to use.

Ideally, all the individual pieces of manure you'll be using should be smaller than a marble. Break up anything marble-sized or larger into smaller chunks. Some people screen their manure based on size, but this isn't really necessary.

Although straw needs to soak before pasteurization, manure does not. It will become appropriately hydrated during the pasteurization process.

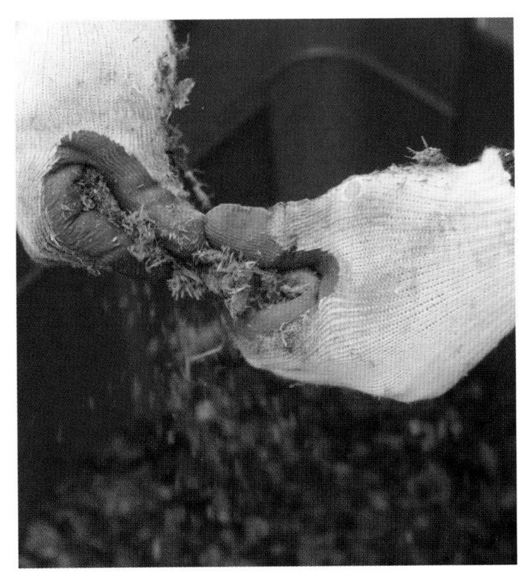

Crumble the manure well before pasteurizing it.

Worm Castings

A final type of manure that can work well in many cases is worm castings. There are several brands of castings made primarily from worms digesting cow manure. This is ideal, but most brands don't list the main ingredient they feed the worms. In many cases it's a peat moss base, which will work, but manure-based castings are preferable.

Mushroom Compost

Mushroom compost, available at garden centers, is not a good substrate for growing mushrooms. This spent compost from large-scale Agaricus mushroom farms works fairly well for gardens, but it's not suitable for growing most species of mushrooms. Most of the nutrients have already been extracted by the previous mushroom growth.

Pasteurizing Small Amounts

If you're working with relatively small amounts of substrate, you have several pasteurization methods to choose from: water bath, steam, and oven pasteurization. I'll explain each method and the substrates for which they're most effective.

Water Bath

A water bath is the method I use most often when pasteurizing small amounts of substrate. It works well in the kitchen for preparing small amounts of straw or manure quickly and reliably.

Water Bath Pasteurization

Materials

- Large pot
- Standoff plate
- Pillowcase
- Thermometer
- Zip ties (optional)
- Brick or other heavy object that can be submerged in water
- Straw or manure

1 **Prepare your substrate.** If you're using straw, cut it into 2" to 3" (5 to 8 cm) pieces (see page 208), and go to step 2. If you're using manure, crumble it into small pieces (see page 211), and go to step 3. If you're using straw, soak it for 1 to 2 hours to ensure that it's properly hydrated before proceeding.

2 **Place your substrate into the pillowcase.** The filled pillowcase should fill the pot at least three-quarters of the way up the side of the pot. Either tie the end of the pillowcase in a knot, or use a zip tie to close it.

3 **Prepare the pot.** Put the standoff plate in the bottom of the pot, and place your pillowcase on it.

4 **Fill the pot with water.** Place the brick or other weight on the pillowcase to keep it submerged.

5 **Heat the water.** Turn on your stove burner and bring the water up to around 155°F (68°C). Keep the temperature between 145 and 165°F (63 and 74°C) for 60 minutes. After 60 minutes, turn off the burner and allow the pot of water to cool. Once the water is cool enough to touch, pull out the pillowcase and let it finish cooling and draining in the sink for several hours. Once it's completely cool, your substrate is ready to use.

Checking Moisture Content

Before spawning, you need to be sure that the moisture content of your substrate is correct. This is fairly easy with straw, as you can simply let all the water drain off. It becomes slightly more complicated with manure and compost, as they tend to hold different amounts of water, depending on the type or brand, and may remain oversaturated even after draining.

I often use a squeeze test to see if the moisture is right. Gently squeeze a handful of the substrate; only a few drops of water should come out. If there's a stream of water, it's oversaturated, and you need to remove more water before spawning. For small amounts of manure, you can simply wring it out with your hands. Certain brands of bagged manure or worm castings become compressed and squirt out of your fingers as mud if you squeeze them. It's probably best to consider steam or oven pasteurization for these substrates.

If you give the substrate a firm squeeze and no water comes out, it may be too dry and you may need to add water. This is rarely a concern for substrates pasteurized with a water bath.

Steam Pasteurization

Steam pasteurizing straw or manure is a viable alternative to hydrating it in a water bath. Many large mushroom farms pasteurize substrate using big steam rooms, or they fill shipping containers with substrate and inject steam into them. Home growers can replicate this process on a smaller scale using a plastic storage bin and a wallpaper steamer. Wallpaper steamers can be found at most home improvement stores and cost between $50 and $100. When selecting a bin, be sure to spend a little extra on one of the thicker models. Steam can compromise the integrity of cheaper bins.

Note that straw should be soaked for 12 to 24 hours and drained before steaming to ensure adequate hydration. Manure and compost should be moistened to near the correct levels before being steamed.

Oven Pasteurization

Cooking manure-based substrates in the oven is a fast and effective method that tends to work best for prebagged substrates and castings. This method has a couple of advantages. First, it's not very messy — no manure water will be splashed around your kitchen. Second, you don't need any special equipment. Finally, most of the nutrients are retained within the substrate, unlike the hot-water bath, which allows some nutrients to leach into the water that's is poured off.

Note that this method is not recommended for pasteurization of straw.

Steam Pasteurization Method

Materials

- Thick plastic storage bin
- Wallpaper steamer
- Heat-resistant caulk or sealant (optional)
- Rubber grommet (optional)
- Manure or straw
- Compost thermometer
- Kitchen timer

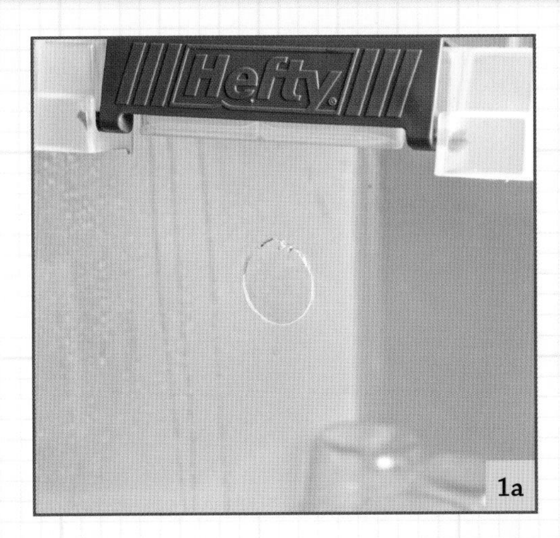

1a

1b

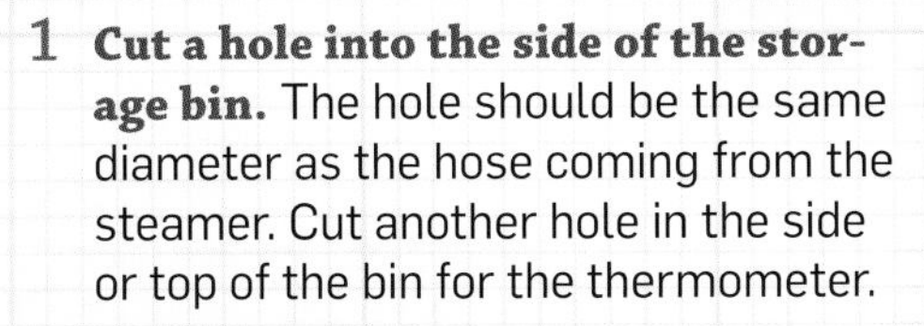

1 **Cut a hole into the side of the storage bin.** The hole should be the same diameter as the hose coming from the steamer. Cut another hole in the side or top of the bin for the thermometer.

2 **Secure the hose.** Remove the large, flat section that spreads the steam onto the wall. The opening on the end of the hose should be exposed. Place the steamer hose just inside the opening of the bin. Seal it with heat-resistant silicone or caulk if the fit isn't very tight. Also consider using a rubber grommet to seal the hose to the bin.

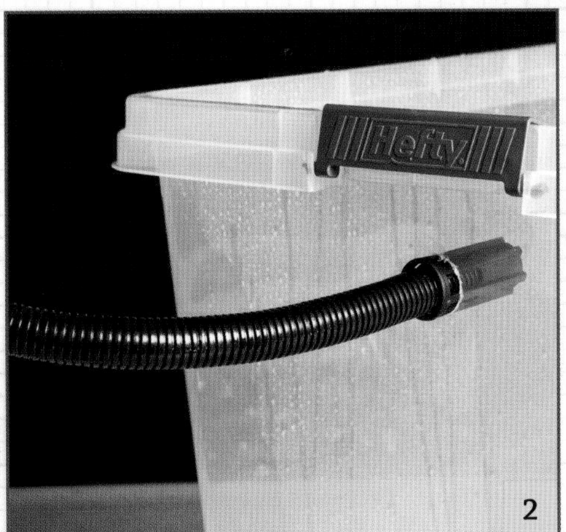

2

3 **Steam the substrate.** You can use a pillowcase to hold the substrate or put it straight into the bin. Slide the compost thermometer into the bin and into the center of your substrate. Fill the steamer with water and turn it on. Once the temperature on the thermometer reaches 145°F (63°C), set the timer for 1 hour.

If the temperature on the thermometer reaches 160°F (71°C), turn off the steamer until the temperature goes down to 150°F (65.5°C). Turn the steamer back on and repeat until 1 full hour has elapsed within the proper temperature range.

When the temperature has stayed at 145 to 150°F (63 to 65.5°C) for 1 full hour, turn off the steamer and allow your substrate to cool in the bin.

Oven Pasteurization Method

Materials

- Aluminum roasting pan, large
- Aluminum foil
- Thermometer
- Water
- Manure-based substrate

1 **Fill the pan with manure.** Prepare your manure by crumbling it appropriately. Wet your manure to just above field capacity (no water should drip from the material unless it's lightly squeezed) and place it in the pan. Ideally, if it's just above field capacity, the evaporation that takes place in the oven will leave it at the right moisture after cooking. Cover the filled pan with aluminum foil.

2 **Heat the oven.** Preheat the oven to the lowest setting, usually between 170 and 250°F (77 and 121°C) for most ovens. Place your thermometer through the foil and into the substrate. Put the pan in the oven. Once the thermometer reads 145°F (63°C), set a timer for 1 hour. Once the thermometer reads 160°F (71°C), turn the oven off. Turn the oven back on once the temperature goes down to 150°F (65.5°C). Repeat this process until one hour has elapsed at the proper temperature. Turn off the oven once the time has elapsed. Allow your substrate to cool in the oven. The substrate can be used when it's completely cool to the touch.

Turkey-roasting pans are a great vessel for oven pasteurizing manure-based substrates.

Pasteurizing in a 55-Gallon Drum

All the pasteurization methods up to this point have been for small amounts of substrate. There are many methods for pasteurizing larger amounts at home; the most common one is described here. For this method, you'll create what is essentially a large pot from a metal drum, and you'll heat water in the drum using a propane burner. This setup is heavy and requires careful consideration of an appropriate location.

To pasteurize straw in a 50-gallon container, start by constructing a basket out of ½-inch wire mesh.

Construct the Basket

Begin by cutting two circles out of chicken wire that are just smaller than the interior diameter of your barrel. These two circles will serve as the base of your interior wire basket. Measure the circumference of the circles you just cut. Cut a length of chicken wire equal to the circumference measurement. This will serve as the sides of your basket. Tie the two cut edges of your side walls together using the thinner metal wire. This should form the circle of your basket. Using thin metal wire, attach the bottom circles to the side wall. You should now have a completed basket. Weave the thicker metal wire through the chicken wire basket. Your basket alone will not be able to support a significant weight. This thicker metal wire becomes the basket's main load-bearing structure and keeps it from falling apart.

Choose the Location

You must consider the best place to locate your drum for cooking. Filled with straw and water, the drum will probably weigh several hundred pounds and cannot be moved. You won't be able to pull the basket out of the drum unaided, either, so you only have a couple of options: you can either create a way to drain the water from the drum, or you can set up a mechanism for pulling the basket out of the water in the drum.

If you install a spigot at the base of the barrel, the water can be drained after the process concludes, removing most of the weight. Be sure to install the spigot in a way that prevents intense heat from affecting it. Don't use silicon or another such heat-sensitive sealant to keep the spigot in place, because it will melt and the drum will leak.

Another option is to set up a ratchet-and-pulley system to pull the basket out of the water once it's finished cooking. To do this, you'll need a high structure that can support the weight of the pulley and wet straw. Be sure to site your setup in a place where you can use a propane burner.

Think carefully about these considerations before you move on. You may have to get creative to find an appropriate location.

Pasteurization

When your basket is constructed and your location found, you can pasteurize your substrate. First place your propane burner in your location. Set one cinder block on either side of your burner. Cinder blocks must be just taller than the surface of the burner, as they will be supporting the weight of the barrel, straw, and water. The burner alone will not support this weight for long.

Place your barrel on the cinder blocks, and put the bricks into the bottom of the barrel to create something similar to the standoff plate of a pressure cooker. Don't put the bricks in the middle of the barrel, as this will inhibit heat transfer while the straw is cooking. Instead, place them closer to the edges.

A large propane-fired burner is used to pasteurize the straw. Place the barrel on cinder blocks surrounding the burner for better support.

Load the basket into the barrel and fill it with straw. Fill the barrel with water until the straw is submerged.

Once the straw has been in the hot-water bath for an hour, remove the basket and allow the straw to cool before inoculating it.

Place the wire basket into the drum. Fill it with straw. Place a weight on top of the straw to keep it submerged.

Fill the barrel with water until all the straw is submerged. Turn on the burner and bring the water up to temperature, as in the hot-water bath method (see page 212). Once 1 hour has elapsed in the proper temperature range, drain the water or remove the straw basket from the drum. Let the straw drain and cool.

Grow Containers

When growing mushrooms on bulk substrates, you have many options for containers. The container you choose depends largely on the species you're growing. Here are some of the most common. Refer to the species pages (see pages 21–24) for additional information.

Plastic Tubing

Polyethylene tubing is most often used to grow Oyster mushrooms on straw. One end of the tubing is sealed off, and straw is packed down into the tube to form an elongated "log" of straw. Appropriate tubing diameters range from 6 to 14 inches (15 to 36 cm). Smaller diameter tubes are prone to drying out; larger diameters prevent the interior section from getting sufficient oxygen and tend not to colonize properly.

Plastic Bins

Most types of mushrooms, other than Oysters, that are grown from bulk substrates are best grown in bin-style containers. This includes *Agaricus* species like White Buttons and Portabellas. The spawn for these species is mixed with the bulk substrate in the bottom of the bin, and covered with a 1- to 2-inch (2.5 to 5 cm) casing layer of moistened vermiculite or peat moss.

When spawning bulk substrates, consider using a spawn-to-substrate ratio between 1:5 and 1:10. The more spawn you use, the faster it will colonize the new substrate. If you try to stretch your spawn to much greater than 1:10, colonization may be so slow that you risk a contaminant setting in before you get the substrate fully colonized. Always remember that growing mushrooms is a race between the mushroom mycelium and every other contaminant in the environment. Give your mushrooms the best conditions for winning the race.

Casing Bulk Substrates

There are several options when dealing with the casing layer with bulk substrates. Most growers allow the mycelium to colonize around 70 percent of the substrate before applying the casing layer. This allows the colonization to be more even across the surface, which should yield a more even pinset once fruiting begins to occur. The spawned substrate can be covered lightly with foil and incubated until it reaches this level of colonization.

Other growers apply a casing layer immediately after the spawn is mixed with the substrate. This also works fine, especially if you have an aggressive species or use a high rate of spawn in your substrate.

Most of the *Agaricus* species do better in thicker substrate layers, so potential growers could also consider plastic bags. Hydroponic stores carry bags that have a black inner surface and a white outer surface, which are great for working with *Agaricus* species. Use them as you would a bin, with the spawned bulk substrate in the base of the bag and a casing layer on the top.

When you work with bulk substrates, you should correlate the thickness of your casing layer to the thickness of your substrate layer. Consider a layer of ½ inch (1.3 cm) for substrate layers up to 2 inches (5 cm) thick, a layer up to 1 inch (2.5 cm) for substrate layers up to 4 or 5 inches (10 or 13 cm) thick, and a casing layer up to 2 inches (5 cm) thick for even larger substrate layers.

Finally, when considering thicker substrate layers, remember that the interior will not be able to colonize if oxygen cannot penetrate into the center. Don't pack down your substrate layer when working with most manure-based substrates, especially if your substrate is made up mostly of small particles with a high moisture content.

Oyster Columns

Oyster mushrooms can be grown from a wide variety of substrates and in a wide variety of containers. The most common way they're grown commercially is from straw packed into polyethylene tubing. Creating columns is probably the most effective method for the home grower as well, and polyethylene tubing is readily available online.

Growing Oyster mushrooms from straw columns has several advantages. The main one is that the straw only has to be pasteurized, not sterilized. You can pasteurize large amounts of straw without expensive special equipment. Another advantage is that you can use premade spawn. If you bought premade spawn for most other mushroom species, you would have to transfer the spawn into a sterilized fruiting substrate under sterile conditions using pressure cookers and a full lab. Oyster mushrooms are one of the only species that you can easily grow without a full lab, as the spawn can reliably be transferred to the fruiting substrate outdoors in the open air.

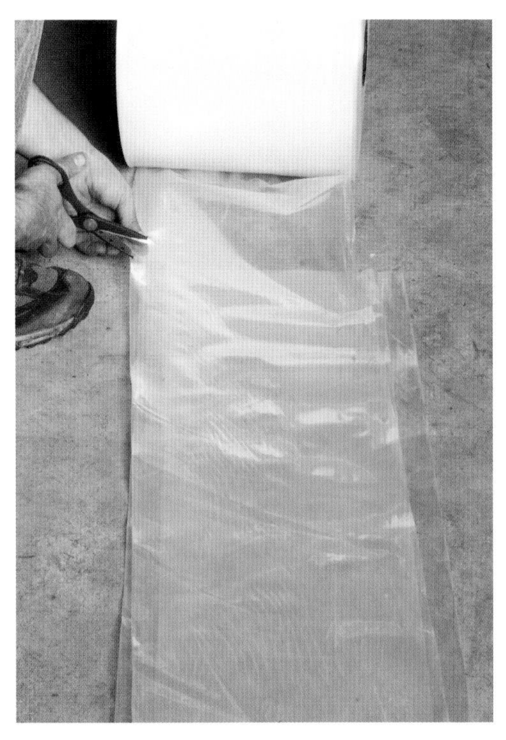

When ordering plastic tubing for your Oyster columns, remember that the measurements given often refer to the "lay flat" diameter — not the diameter of the tubing once it's open and filled.

Selecting Tubing

Most growers use 4-mil plastic tubing. Keep in mind that tubing diameter often refers to the "lay flat" diameter, not the expanded-circle diameter. This means that if you were to order 8-inch- (20 cm) diameter tubing, it would arrive as a roll of plastic that is 8 inches (20 cm) wide. Once you cut off your desired length of tubing and open it up, it would create a log that is just over 5 inches (13 cm) in diameter.

The optimum opened diameter for Oyster columns is between 6 and 12 inches (15 and 30 cm). If you try to create thinner columns, the logs will tend to dry out more easily and the harvest won't be as great. If you attempt to use columns greater than 12 inches (30 cm), it becomes difficult for the substrate within the interior of the log to fully colonize. Oxygen won't be able to reach this interior portion, and the mycelium won't be able to colonize it effectively. You'll also see diminishing returns for logs that are more than 10 inches (25 cm) in diameter. Instead of increasing the diameter of the column, you'd have greater success starting a new, smaller diameter column with the spawn and straw.

Lay-Flat Diameter	Circumference	Opened Diameter
8" (20 cm)	16" (41 cm)	5.10" (13 cm)
10" (25 cm)	20" (51 cm)	6.37" (16 cm)
12" (30 cm)	24" (61 cm)	7.64" (19 cm)
14" (36 cm)	28" (71 cm)	8.92" (23 cm)
16" (41 cm)	32" (81 cm)	10.19" (26 cm)
18" (46 cm)	36" (91 cm)	11.46" (29 cm)
20" (51 cm)	40" (102 cm)	12.74" (32 cm)

Length of Columns

The best length for Oyster columns depends on the grower. I tend to make the individual columns about 3 feet (1 m) long, and then hang a couple of them, one on top of the other, ultimately creating a 6-foot (2 m) column. I find that these shorter columns are easier to create and move around. Other growers tend to create single logs that are 6 feet (2 m) long. The ideal length for yield and workability probably lies somewhere within this range.

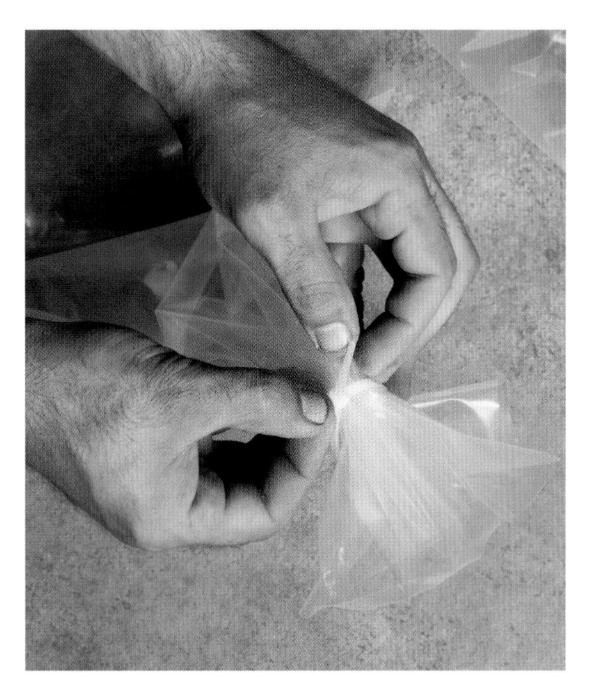

Seal the ends of columns securely with zip ties.

Sealing the Ends

Seal the ends of the columns with zip ties, which are relatively inexpensive and very effective. If you pull them tight, they're very unlikely to slip off the end of the tubing while it's being made or fruited. To hang two columns together, I buy special zip ties with a plastic loop on the end and attach the lower column to the upper one with an S hook.

You can also seal the ends of the tubes with twine, string, tape, or anything that's not going to slip off or break and allow the straw to fall out the bottom. Remember to leave a little bit of extra securing material at the ends so there's room to seal the columns without inhibiting the amount of available space you have for fruiting.

Preparing the Substrate

After you pasteurize your straw (see page 218), let it cool down completely before proceeding. When the straw is no longer warm to the touch, it's time to mix the spawn into the substrate and create your column. First, break up the spawn in the spawn bag as much as possible, leaving no chunks or clumps. The more you break it up, the more substrate you'll be able to inoculate with the spawn.

Different growers take different approaches to the next phase. Some growers like to lay the pasteurized straw out on a table, and spread the spawn over the straw before packing it into the tubes. This method probably makes the best use of spawn, as you can spread out all of your spawn fairly equally across the amount of available straw. I prefer a slightly different method: I place a handful of straw into the tube, followed by a handful of spawn, and repeat this process until the column is full. Either way will work fine, so choose what works best for your situation.

Sawdust spawn is a great choice for inoculating straw. Crumble the spawn over the pasteurized straw before loading it into the columns.

How Much Spawn?

The first obvious question is how much spawn to use per given amount of straw. I've encountered several different ratios for spawning straw, and I'm not enthusiastic about any of them. The first is 2 to 5 percent of the wet weight mass, but how many of you are going to weigh out the straw and the spawn? The second ratio is one 5-pound (2.3 kg) bag of spawn per bale of straw. This one is a little easier to break down, but home growers will probably not be pasteurizing entire bales at once, or even significant portions of a bale. Instead, I use a hand measurement. I add one healthy handful of straw into my container, and follow it up with a small handful of spawn, repeating the process until the container is full. There are no disadvantages to using too much spawn, especially if you're making your own. Excess spawn will simply speed up the colonization process and give you mushrooms more quickly; however, not using enough spawn can prevent the substrate from becoming fully colonized.

When you start spawning straw, try making a series of tubes with different amounts of spawn in each. Keep the size of your handful of straw consistent, but change the size of your handful of spawn, or try adding set measures of spawn to each handful of straw. If you label all these tubes and keep track of how they perform, you'll be able to adjust your levels of spawn in the future to make the most of what you have available.

Packing the Columns

No matter what type of container you use, it's best to pack the straw very tightly into it. The more you pack the straw, the closer the individual stalks of cut straw will be to each other. This will make it easier for the growing mycelium to jump from

Pack the straw tightly into the column. The finished column should have little or no give when it's squeezed.

stalk to stalk, and will ultimately mean faster colonization. It also allows a greater quantity of substrate in each tube, which often means greater yields per volume of columns.

When I say I pack the straw tightly, I mean as tightly as I can. Every couple of times I add straw, I push my fist firmly down into the back to compress the substrate. How firmly you need to pack the straw depends on how small you cut it. If the straw is uncut, pack it as firmly as you can with a quick push. If the straw is cut into smaller pieces, it's much more likely to compress naturally, so it may not require quite as much strength. Your finished column should not have much give to it when squeezed. If the walls of the bag have much flex, you haven't packed the straw tightly enough. The bag doesn't need to be rock hard to the touch, but it should be fairly firm.

Puncture the column to allow oxygen to enter and to give the mushrooms space to fruit.

Pressing Holes

After your bags are made, you'll need to cut holes in them to allow oxygen to penetrate into your column and give your mushrooms space to fruit. Make the holes soon after you make up the tube, or bacterial contamination may set in and inhibit the future growth of your spawn.

Most growers use four-sided arrowheads to poke X-shaped holes into the bag. Multiple arrowheads can be mounted on a piece of wood to speed up the process. The slits that are created should be spaced roughly 4 inches (10 cm) apart, and each new row should be offset to form a diamond pattern with the previous row. Each slit should be more than ⅛ inch (3 mm) across, but no more than 1 inch (2.5 cm) across. Most of my arrowhead punctures are usually around the upper portion of that range.

Incubation

After the holes are created, incubate the columns for a week or so to allow them to colonize. Not all growers incubate Oyster columns; some growers move them directly from inoculation into the fruiting room. If you have room in your fruiting environment, you can go in either direction. It can take between 10 and 14 days for the primordia to begin to form on the columns, so you have the option of incubating for one week or not.

Fruiting and Harvest

Oyster mushrooms tend to fruit in large clusters. Once you see the primordia begin to form, it should only take 3 to 7 days before they reach full maturity and are ready to harvest. It's best to harvest the mushrooms before they begin producing spores. You'll know the mushrooms are producing spores when you begin to see a white powder on mushrooms near the bottom of the cluster. Another sign that Oyster mushrooms are ready to harvest is when the outer edge of the cap begins to turn up.

You can harvest the mushrooms by grasping the clusters at the base and twisting the entire cluster off the column at once. Refrigerate the mushrooms immediately after harvest.

Large amounts of Oyster mushrooms can be grown from columns of inoculated straw.

Consecutive Flushes

After you harvest all the mushrooms on the column, give the entire structure a good misting. If you maintain fruiting conditions, you should receive another harvest in about 2 weeks. Depending on the diameter of your column and how much moisture it's able to maintain, you can expect three to five harvests from each column.

Spore Load

Oyster mushrooms tend to produce significantly more spores than most other species. If you're growing and harvesting for an extended period of time, consider wearing some type of mask while working in the growroom. Exposure to large amounts of spores over long periods have been known to give some people respiratory problems.

Fungus Gnats

Another common problem with Oyster mushrooms is the fungus gnat, a pest who is often attracted to Oyster mushroom mycelium. It may be necessary to add fly traps to your growroom if they become a problem. Oyster mushrooms are particularly sensitive to chemicals, and malformed growth may result from the use of common insecticides.

If gnats become a problem, completely clear out your growroom and clean it thoroughly before you bring more fresh substrate into it. This should help remove many of the pests before they begin growing in your new substrate.

RESOURCES

Online Mushroom Resources

Magnificent Mushrooms
www.magnificentmushrooms.com
Supplies

The Mushroom Farm
www.mushroomfarm.com/forum
Get your questions answered here.

Mycotopia
http://mycotopia.net

Shroomery
www.shroomery.org

North American Mycological Association
www.namyco.org

Wild Mushrooms

The Hoosier Mushroom Society
www.hoosiermushrooms.org

Mushroom Expert
www.mushroomexpert.com

Mushroom Observer
www.mushroomobserver.org

Morel Hunters
www.morelhunters.com
Post your morel photos to the morel progression map.

Further Reading

Stamets, Paul. *MycoMedicinals.* MycoMedia, 1999. For a great summary of medicinal mushrooms and their benefits.

_____. *The Mushroom Cultivator.* Agarikon Press, 1984.

_____. *Growing Gourmet and Medicinal Mushrooms.* Ten Speed Press, 2000.

INDEX

Page numbers in *italic* indicate photos or illustrations: those in **bold** indicate charts.

A

B